T0221101

Annals of Mathematics Studies

Number 114

THE WILLIAM H. ROEVER LECTURES IN GEOMETRY

The William H. Roever Lectures in Geometry were established in 1982 by his sons William A. and Frederick H. Roever, and members of their families, as a lasting memorial to their father, and as a continuing source of strength for the department of mathematics at Washington University, which owes so much to his long career.

After receiving a B.S. in Mechanical Engineering from Washington University in 1897, William H. Roever studied mathematics at Harvard University, where he received the Ph.D. in 1906. After two years of teaching at the Massachusetts Institute of Technology, he returned to Washington University in 1908. There he spent his entire career, serving as chairman of the Department of Mathematics and Astronomy from 1932 until his retirement in 1945.

Professor Roever published over 40 articles and several books, nearly all in his specialty, descriptive geometry. He served on the council of the American Mathematical Society and on the editorial board of the Mathematical Association of America and was a member of the mathematical societies of Italy and Germany. His rich and fruitful professional life remains an important example to his Department.

DIFFERENTIAL SYSTEMS AND ISOMETRIC EMBEDDINGS

BY

PHILLIP A. GRIFFITHS AND GARY R. JENSEN

THE WILLIAM H. ROEVER LECTURES
IN GEOMETRY
WASHINGTON UNIVERSITY
ST. LOUIS

PRINCETON UNIVERSITY PRESS

PRINCETON, NEW JERSEY

1987

Copyright © 1987 by Princeton University Press
ALL RIGHTS RESERVED

The Annals of Mathematics Studies are edited by
William Browder, Robert P. Langlands, John Milnor, and Elias M. Stein
Corresponding editors:
Stefan Hildebrandt, H. Blaine Lawson, Louis Nirenberg, and David Vogan

Clothbound editions of Princeton University Press books are printed on acid-free paper, and binding materials are chosen for strength and durability. Paperbacks, while satisfactory for personal collections, are not usually suitable for library rebinding

ISBN 0-691-08429-7 (cloth)
ISBN 0-691-08430-0 (paper)

Printed in the United States of America
by Princeton University Press, 41 William Street
Princeton, New Jersey

☆

Library of Congress Cataloging in Publication data will
be found on the last printed page of this book

William H. Roever 1874–1951

Contents

PREFACE

This monograph is an elaboration of the William H. Roever Lectures delivered by Phillip Griffiths in January, 1984, at Washington University in St. Louis. It contains an exposition of the theory of quasi-linear Pfaffian differential systems, with particular emphasis on the theory of the characteristic variety. The theory is illustrated throughout with a detailed application to the problem of the local isometric embedding of a Riemannian manifold into Euclidean space, where the theory quite directly and naturally uncovers the essential pointwise algebraic conditions necessary for the existence of such embeddings. In the last chapter the theory is applied to the problem of embedding Cauchy-Riemann structures.

We wish to thank Micki Wilderspin for her excellent job of typing and for her efficient organization in keeping the manuscript in order as it travelled in bits and pieces between St. Louis and Durham.

Phillip Griffiths
Durham, North Carolina

Gary R. Jensen
St. Louis, Missouri

November, 1984

COMMONLY USED NOTATION

1) $(\overline{X}, \overline{ds}^2)$ is an abstract Riemannian manifold.

2) $\mathbf{E}^N$ is Euclidean N-space.

3) $x: \overline{X} \to X \subset \mathbf{E}^N$ is an isometric embedding with image X.

4) $y \in \overline{X}$ is a point with tangent space $T_y\overline{X}$ and cotangent space $T_y^*\overline{X}$.

5) For $X \subset \mathbf{E}^N$ we denote by $x \in X$ a point with tangent space T_xX, normal space N_xX, and 2nd fundamental form $II(x) \in S^2T_x^*X \otimes N_xX$.

6) S^qV denotes the qth symmetric product of a vector space V.

7) Λ^qM denotes the C^∞ q-forms on a manifold M, and $\Lambda^*M = \bigoplus_{q \geq 0} \Lambda^qM$. We also set $C^\infty M = \Lambda^0M$.

8) A Pfaffian differential system (I,J) on M is given by sub-bundles $I \subset J \subset T^*M$.

9) $\underline{I} \subset \Lambda^1M$ is the space of C^∞ sections of $I \subset T^*M$.

10) $\underline{I}$ is locally generated over $C^\infty M$ by $\theta^1, \ldots, \theta^s$; $\underline{J}$ is locally generated over $C^\infty M$ by $\theta^1, \ldots, \theta^s, \omega^1, \ldots, \omega^p$.

11) $\{\underline{I}\} \subset \Lambda^*M$ is the algebraic ideal generated by $\underline{I}$; sometimes we locally write $\{\underline{I}\} = \{\theta\}$ where θ is thought of as the collection of θ^α.

12) $\mathcal{I} \subset \Lambda^*M$ is the differential ideal generated by $\underline{I}$.

CONVENTIONS

1) Summation convention, meaning sum all pairs of repeated indices in a product, will be used throughout except where explicitly stated otherwise.

2) The following ranges of indices will be used

$$1 \leq a,b,c \leq N$$
$$1 \leq i,j,k \leq n$$
$$n+1 \leq \mu,\upsilon \leq N$$
$$1 \leq \alpha,\beta \leq s \qquad s = \text{rank } I$$
$$1 \leq \rho,\sigma \leq p \qquad p = \text{rank } J/I$$
$$0 \leq A,B,C \leq n+1$$

3) (4.19) refers to statement on equation (19) in Section 4.

4) (19) refers to statement on equation (19) in the same section.

DIFFERENTIAL SYSTEMS AND ISOMETRIC EMBEDDINGS

CHAPTER 1

INTRODUCTION

Let $\overline{X}$ be an n-dimensional Riemannian manifold with metric locally given by

$$ds^2 = g_{ij}(y)dy^i dy^j$$

where $y = (y^1,\ldots,y^n)$ are local coordinates on $\overline{X}$. We denote by E^N Euclidean N-space with points $x = (x^1,\ldots,x^N)$ and standard flat metric

$$\langle dx, dx\rangle = \sum_a (dx^a)^2.$$

By an <u>isometric embedding</u> of $\overline{X}$ in E^N we shall mean a one-to-one C^∞ mapping

$$x : \overline{X} \to E^N, \tag{0}$$

such that the differential (denoted here by dx instead of x_*)

$$dx : T_y\overline{X} \to T_{x(y)}E^N$$

is an isometry for each point $y \in \overline{X}$. Equivalently,

$$(1) \qquad \langle dx(y), dx(y) \rangle = \overline{ds}^2,$$

or in local coordinates

$$(2) \qquad \frac{\partial x^a(y)}{\partial y^i} \frac{\partial x^a(y)}{\partial y^j} = g_{ij}(y).$$

We set

$$x(\overline{X}) = X \subset E^N, \quad \text{and}$$
$$N = n + r \quad \text{where} \quad r = \text{codimension of} \quad x$$

and define the <u>embedding dimension</u> to be

$$N(n) = n(n + 1)/2 = n + n(n - 1)/2.$$

According to (2), isometric embeddings are given at least locally by solutions to a non-linear, 1st order P.D.E. system that is overdetermined if $N < N(n)$, determined if $N = N(n)$, and underdetermined if $N > N(n)$. The same is true globally if we talk only of isometric immersions.

A classical problem is to study the existence and uniqueness of isometric embeddings. There is both a local question and a global one. Even though the global problem is clearly the one of ultimate interest, we shall in these talks be concerned entirely with the local question. Our goal will be to illustrate how some recent developments in the theory of exterior differential systems, especially the theory of characteristic varieties for such systems, may be used to shed interesting new light on even local isometric embeddings.

A few comments concerning the system (2) may be helpful. We restrict to the determined case $N = N(n)$. Geometrically, one expects that the Riemannian curvature R of $(\overline{X},\overline{ds^2})$ should enter into the picture. Since

$$R = R(g,\partial g,\partial^2 g)$$

involves up through 2nd derivatives of $g_{ij}(y)$, this will only happen explicitly if we differentiate (2) twice. In intrinsic terms, we must prolong twice the relevant exterior differential system. But when this is done the resulting P.D.E. system appears to be overdeterminted. Morally we know this isn't so - the precise mathematical way of dealing with this involves the involutivity of the exterior differential system. The notions of prolongation and involution are two of the most subtle concepts in the theory of exterior differential systems. Part of what we shall try to do in these talks is to explain them by carefully working through the isometric embedding system.

Before outlining specifically what we shall try to do, let us briefly recall a few highlights of the classical theory (a slightly more detailed history is given in Berger-Bryant-Griffiths [1]). Concerning existence the main local results are the following:

(3) THEOREM (Cartan-Janet). In the real-analytic case there exist local isometric embeddings

$$x:\overline{X}^n \to E^{N(n)}.$$

Recall that these are the dimensions where the P.D.E. system (2) is determined, so that Cartan-Janet is the result that one would like to have.

(4) THEOREM (Classical - cf. Spivak [1]). There exist local C^∞ isometric embeddings

$$x:\overline{X}^2 \to E^3$$

is case the Gaussian curvature $K \neq 0$.

The problem turns out to be elliptic if $K > 0$ and hyperbolic if $K < 0$. Just recently Lin [1] has made important progress on the case when K changes sign.

(5) THEOREM (Nash-Greene; cf. Greene [1]). There exist local C^∞ isometric embeddings

$$\overline{X}^n \to E^{N(n)+n}.$$

The method of proof is to apply an iteration scheme analogous to Newton's methods for finding roots of a polynomial. The dimension restriction comes from the fact that at each stage the linear problem to be solved turns out to be a system of algebraic equations. Even so, the iteration scheme "loses two derivatives" at each step, and consequently it is necessary at each stage to apply smoothing operators. This is the genesis of the famous Nash-Moser implicit function theorem (cf. Hamilton [1] for a recent overall account).

We emphasize that we are ignoring the global C^∞ theory of Nash-Gromov; a survey of these results may be found in Greene [1], Gromov-Rokhlin [1], or Gromov [1].

Before discussing uniqueness we introduce a little terminology. Given a submanifold

$$X^n \subset E^{n+r}, \tag{6}$$

at each point $x \in X$ there is defined the

second fundamental form

$$II(x) \in N_xX \otimes S^2T_x^*X$$

where N_xX is the normal space to X at x and $S^2T_x^*X$ is the 2nd symmetric power of the cotangent space. For each n,r there will be a Zariski open subset (to be specified in definition (6.64))

$$U_{n,r} \subset \mathbb{R}^r \otimes S^2\mathbb{R}^n$$

in the space of $\mathbb{R}^r$-valued quadratic forms, and we say that the submanifold (6) is non-degenerate in case $II(x) \in U_{n,r}$ for each $x \in X$. It will be seen that this has intrinsic meaning in the sense that $U_{n,r}$ is invariant under $O(r) \times O(n)$ acting on $\mathbb{R}^r \otimes S^2\mathbb{R}^n$. An isometric embedding will be said to be non-degenerate in case the image manifold is non-degenerate. This may be thought of as follows. If we set (here "D" stand for degenerate)

$$D_{n,r} = \mathbb{R}^r \otimes S^2\mathbb{R}^n \setminus U_{n,r},$$

then those isometric embeddings with

$$II(x) \in D_{n,r}$$

satisfy an additional 2nd order P.D.E. system beyond (2). Thus, roughly speaking, non-degeneracy for an isometric embedding means that no additional equations beyond (2) are imposed. In the first six chapters of these talks we shall consider only non-degenerate isometric embeddings.

For these embeddings, we shall say that (0) is rigid in case the mapping x is uniquely determined, up to a rigid motion of E^N, by the $\overline{ds}^2$ on $\overline{X}$. The main classical local uniqueness result is the following

(7) THEOREM (Allendoerfer-Beez) A generic local isometric embedding is rigid in case

$$r \leq [n/3].$$

Here generic means: 1) the first normal space of the submanfold has dimension r at every point; and 2) the submanifold has maximal type, (see the following references). A proof of this result may be found in Chern-Osserman [1] or Kobayashi-Nomizu [1], or Spivak [1]; related rigidity questions are discussed in Kaneda-Tanaka [1].

In these talks we will study isometric embeddings using exterior differential systems (E.D.S.'s). If $\mathcal{F}(\overline{X})$ denotes the bundle of orthonormal frames on the

Riemannian manifold $\overline{X}$, then we shall set up an E.D.S. on

$$\mathcal{F}(\overline{X}) \times \mathcal{F}(E^N)$$

whose local integral manifolds (i.e., "solutions") are all Darboux framings of local isometric embeddings (0). The approach is basically the same as the classical one of Cartan [1], except that we will prolong the system differently. Our main technique is to apply the theory of the characteristic variety of an E.D.S. (cf. Bryant et al. [1]) to the isometric embedding problem, leading to the following (cf. Berger-Bryant-Griffiths [2] and Bryant-Griffiths-Yang [1]):

(8) THEOREM: a) A local isometric embedding is rigid if

$$r \leq n, \quad \text{when} \quad n \geq 8;$$

$$r \leq n - 1, \quad \text{when} \quad n = 4,\ 5 \quad \text{or} \quad 7$$

$$r \leq 4, \quad \text{when} \quad n = 6.$$

b) It depends only on constants (to be explained below) if

$$r \leq (n - 1)(n - 2)/2 = n(n - 1)/2 - (n - 1).$$

c) If $\det(R_{ij}) \neq 0$, then local C^{∞} embeddings

$$\bar{X}^3 \subset E^6$$

exist in the case $n = 3$.

Our results will also give a pretty good picture of the characteristic variety for arbitrary n, so that the general C^{∞} local isometric embedding problem is in some sense reduced to a question in linear P.D.E. theory (in fact, to solving with suitable estimates the generic 1st order determined linear P.D.E. system; cf. Bryant-Griffiths-Yang [1]).

As a further illustration of the use of exterior differential systems and their characteristic varieties, in Section 7 we will study local isometric embeddings (0) in case $(\bar{X}, \overline{ds}^2)$ is a Riemannian manifold of constant sectional curvature $k \leq 0$. With the case $k = -1$ in mind we first investigate the situation when $k = 0$, $N = 2n$, and the embedding is required to be suitably non-degenerate. The relevant algebra, which is due to E. Cartan, is then easily adaptable to the case $k = -1$, $N = 2n - 1$ (this turns out to be the minimal N). Both situations are beautiful examples of overdetermined, hyperbolic exterior differential systems.

As indicated, our philosophy is to study isometric embeddings via the general theory of characteristic varieties of differential systems. As is becoming increasingly clear, this is a powerful method for studying certain non-linear P.D.E. systems that arise naturally in geometric problems (cf. the examples in Bryant-et al. [1]). It is our feeling that what has been done so far is the tip of an iceberg; the story told by these talks hopefully may be viewed as only the beginning chapter in an extensive development. In particular, we would like to mention the two main general areas that seem to us ripe for study:

i) Use of the complex characteristic variety to study naturally arising elliptic systems (other than those of finite type).

ii) Use of the characteristic variety in global questions.

Regarding the latter, we mention the following problems:

I. Are there curvature assumptions on a compact $(\overline{X}^3, \overline{ds}^2)$ that guarantee rigidity of an isometric embedding

$$\overline{X} \to E^6?$$

Here we have in mind generalizing the famous Cohn-Vossen Theorem. (See Chern [1] or Spivak [1]).

There is the following result of Chern-Kuiper [1]: A compact $(\overline{X}^n, \overline{ds}^2)$ with non-positive sectional curvature cannot be isometrically immersed into E^{2n-1}. (See Kobayashi-Nomiza [1]). In Chapter 7 we prove that a flat $(\overline{X}^n, \overline{ds}^2)$ cannot be isometrically embedded in a non-degenerate fashion even locally in E^{2n-1}, while a space $(\overline{X}^n, \overline{ds}^2)$ with constant negative sectional curvature can be locally isometrically non-degenerately embedded in E^{2n-1}. The following problem is open for $n \geq 3$:

II. Does there exist a global isometric embedding

$$H^n \to E^{2n-1},$$

where H^n is the n-dimensional hyperbolic space form? Here we obviously have in mind the classical Hilbert Theorem. (See Do Carmo [1]).

III. Are there curvature assumptions on $(\overline{X}^3, \overline{ds}^2)$ that prevent the existence of a global isometric embedding

$$\overline{X} \to E^6?$$

Here, we again have in mind the Hilbert Theorem and its extension by Efimov [1].

CHAPTER 2

STRUCTURE EQUATIONS OF $X^n \subset \mathbf{E}^N$

Throughout these talks we shall use without comment the calculus of differential forms. Moreover, we shall frequently omit pullback and restriction notation; as most of you know, if one keeps in mind the geometric picture then it is possible to use differential forms accurately and effectively without unduly worrying about where they are defined. We shall also use summation convention, and shall employ the following ranges of indices

$$
\begin{aligned}
&1 \leq a,b,c \leq N = n + r \\
(1) \qquad &1 \leq i,j,k \leq n \\
&n + 1 \leq \mu,\nu \leq n + r.
\end{aligned}
$$

Orthonormal frames on $\mathbf{E}^N$ will be denoted by $(x, e_1, \ldots, e_N)$, sometimes abbreviated to $(x; e_a)$. The set of all frames constitutes the orthonormal frame bundle

$$\mathcal{F}(\mathbf{E}^N) \overset{x}{\to} \mathbf{E}^N$$

with fibre $O(N)$. Upon choice of a reference frame, $\mathcal{F}(\mathbf{E}^N)$ may be identified with the group $E(N)$ of rigid motions of $\mathbf{E}^N$.

It is important to make this identification explicit. To do this we represent $E(N)$ in $GL(N+1;\mathbb{R})$ by

$$E(N) = \{\begin{bmatrix} A & a \\ 0 & 1 \end{bmatrix} : A \in O(N),\ a \in \mathbb{R}^N\}.$$

The standard action of $GL(N+1;\mathbb{R})$ on $\mathbf{E}^{N+1}$ induces the action of $E(N)$ on

$$\mathbf{E}^N = \left\{\begin{bmatrix} x \\ 1 \end{bmatrix} : x \in \mathbf{E}^N\right\} \subseteqq \mathbf{E}^{N+1}.$$

For simplicity we write (a,A) for $\begin{bmatrix} A & a \\ 0 & 1 \end{bmatrix}$. Then the multiplication rule in $E(N)$ is

$$(a,A)(b,B) = (Ab + a, AB),$$

and the action of $E(N)$ on $\mathbf{E}^N$ is

$$(a,A)x = a + Ax.$$

For a reference frame of $\mathbf{E}^N$ we choose $(0,\epsilon_1,\ldots,\epsilon_N)$, where $0 \in \mathbf{E}^N$ is the origin and $\epsilon_1,\ldots,\epsilon_N$ is the standard basis of $\mathbb{R}^N$. The identification is then given by the bundle isomorphism

$$E(N) \cong \mathcal{F}(\mathbf{E}^N),$$
$$(a,A) \to (a,A)_*(0,\epsilon_1,\ldots,\epsilon_N) = (a,e_1,\ldots,e_N)$$

where $e_i = A_*\epsilon_i$ is the ith column of A.

The Lie algebra $\mathcal{E}(N)$ of $E(N)$ is represented in $\mathcal{G}\ell(N+1;\mathbb{R})$ by

$$\mathcal{E}(N) = \left\{\begin{vmatrix} X & x \\ 0 & 0 \end{vmatrix} : X \in \mathfrak{o}(N),\ x \in \mathbb{R}^N\right\}.$$

We write (x,X) for $\begin{vmatrix} X & x \\ 0 & 0 \end{vmatrix}$. The bracket operation is

$$[(x,X),(y,Y)] = (Xy - Yx, XY - YX).$$

The Maurer-Cartan form Φ of $E(N)$ is just the Maurer-Cartan form of $GL(N+1;\mathbb{R})$ restricted to $E(N)$. Thus, at $(a,A) \in E(N)$,

$$\Phi_{(a,A)} = (a,A)^{-1}d(a,A) = \begin{vmatrix} A^{-1} & -A^{-1}a \\ 0 & 1 \end{vmatrix} \begin{vmatrix} dA & da \\ 0 & 0 \end{vmatrix}$$

$$= \begin{pmatrix} A^{-1}dA & A^{-1}da \\ 0 & 0 \end{pmatrix} = (A^{-1}da, A^{-1}dA) \in \mathcal{E}(N).$$

If we set

$$(2) \qquad \text{(i)} \quad (\omega^a) = A^{-1}da$$

$$\text{(ii)} \quad (\omega^a_b) = A^{-1}dA,$$

then (ω^a) is a left-invariant $\mathbb{R}^N$-valued 1-form on $E(N)$, (ω^a_b) is a left-invariant $o(N)$-valued 1-form on $E(N)$, and

$$\phi = ((\omega^a),(\omega^a_b)).$$

If we let e_a denote the ath column of $A \in O(N)$, and if we multiply the equations in (2) by A, then we obtain an equivalent formulation

$$(2) \qquad \text{(i)} \quad da = \omega^a e_a$$

$$\text{(ii)} \quad de_a = \omega^b_a e_b.$$

The left sides of (2) mean simply the differentials of the column vectors $a, e_a \in \mathbb{R}^N$.

One calculates easily, using the rule for the derivative of the inverse of a matrix, that

$$d\phi = -\phi\wedge\phi,$$

which, when written out in detail, gives the

Maurer-Cartan, or structure, equations of $E(N)$

(3)
$$\text{(i)}\quad d\omega^a = -\omega^a_b \wedge \omega^b$$
$$\text{(ii)}\quad d\omega^a_b = -\omega^a_c \wedge \omega^c_b.$$

Under our identification $\mathcal{F}(\mathbf{E}^N) = E(N)$ the projection $x:E(N) \to \mathbf{E}^N$ is given by $x(a,A) = a$. The pull-back by x of the standard Riemannian metric on $\mathbf{E}^N$ is given by

$$(dx,dx) = \sum_a (\omega^a)^2.$$

A moving frame on an open subset $U \subseteq E^N$ is a section $(x,e):U \to E(N)$. Thus $x:U \to \mathbb{R}^n$ is the identity map, and $e:U \to O(N)$ so that the columns e_a of e are maps $e_a:U \to \mathbb{R}^N$.

The fibre of $x:E(N) \to \mathbf{E}^N$ over any $a \in \mathbf{E}^N$ is

$$x^{-1}\{a\} = \{(a,A):A \in O(N)\}.$$

Thus for a constant, we have

$$(a,A)^{-1}d(a,A) = (0,A^{-1}dA),$$

which means that the tangent space to the fibre through

(a,A) is defined by the equations

$$\omega^a = 0.$$

Given a submanifold $X^n \subset \mathbf{E}^{n+r}$ a <u>Darboux frame</u>, also called an adapted frame, is a frame $(x, e_1, \ldots, e_n, e_{n+1}, \ldots, e_{n+r})$, sometimes abbreviated $(x; e_i; e_\mu)$, satisfying

$$\begin{aligned} & x \in X \\ & e_i \in T_x X \quad \text{is tangent to } X \\ & e_\mu \in N_x X \quad \text{is normal to } X. \end{aligned}$$

The set of Darboux frames forms a bundle

$$\mathcal{F}(X) \to X$$

with fibre $O(n) \times O(r)$ (spinning the tangent and normal frames). Since

$$\dim \mathcal{F}(\mathbf{E}^N) - \dim \mathcal{F}(X) = r + nr$$

we expect the sub-bundle

$$T\mathcal{F}(X) \subset T\mathcal{F}(\mathbf{E}^N)|_{\mathcal{F}(X)}$$

to have its fibre defined by $r + nr$ linear equations expressed in terms of the ω^a and ω^a_b. We shall now determine these equations.

The basic observation is that the restriction, (i.e., pull-back by the inclusion map),

$$\omega^\mu \Big|_{\mathcal{F}(X)} = 0. \tag{4}$$

The intuitive reason for this is that, when $(x;e_a)$ moves infinitesimally in $\mathcal{F}(X)$ to $(x + \delta^a e_a, e_a + \delta^b_a e_b)$, then to first order $x + \delta^a e_a$ lies in X implies that $\delta^\mu = 0$, which is just (4). To verify this formally it suffices to show that for every local section $(x;e) = (x;e_i,e_\mu)$ of $\mathcal{F}(X) \to X$

$$(x;e)^* \omega^\mu = 0,$$

(because the ω^μ already annihilate any vertical vector). But, regarding $(x;e)$ as a map into $\mathcal{F}(E^N) \supseteq \mathcal{F}(X)$, we have

$$(x;e)^*(\omega^a,\omega^a_b) = (x;e)^*\Phi = (e^{-1}dx, e^{-1}de),$$

while $x \circ (x,e) = x$ lies in X implies that

$$dx = x_* \circ (dx,de) \in T_x X = \mathrm{span}(e_i).$$

Combining these we have

$$dx = ((x;e)^*\omega^a)e_a \in \operatorname{span}(e_i),$$

and thus $(x;e)^*\omega^\mu = 0$.

Using terminology to be explained later, we have that $\mathcal{F}(X)$ is an integral manifold of the exterior differential system (E.D.S.)

$$(5) \quad \begin{aligned} &\text{(i)} \quad \omega^\mu = 0 \\ &\text{(ii)} \quad \bigwedge_i \omega^i \wedge \bigwedge_{i<j} \omega^i_j \wedge \bigwedge_{\mu<\nu} \omega^\mu_\nu \neq 0 \end{aligned}$$

on $\mathcal{F}(E^N)$, (we shall see below that (ii) is satisfied). From (4) we conclude that the induced metric on X is given (up on $\mathcal{F}(X)$) by the quadratic differential form

$$I = \Sigma(\omega^i)^2.$$

More substantially, using the fact that "d" commutes with restriction, from (4) we have

$$0 = d\omega^\mu \Big|_{\mathcal{F}(X)} = \omega^\mu_j \wedge \omega^j + \omega^\mu_\nu \wedge \omega^\nu \Big|_{\mathcal{F}(X)} = \omega^\mu_j \wedge \omega^j \Big|_{\mathcal{F}(X)}.$$

From now on we shall omit the cumbersome restriction signs, so that we have on $\mathcal{F}(X)$

$$\text{(6)}\qquad \begin{aligned} &\text{(i)} \quad \omega^{\mu} = 0 \\ &\text{(ii)} \quad \omega^{\mu}_{j} \wedge \omega^{j} = 0. \end{aligned}$$

At this juncture one traditionally invokes the

(7) CARTAN LEMMA: Let $\{\varphi_i\}$ be everywhere linearly independent 1-forms on a manifold, and let $\{\psi_i\}$ be 1-forms satisfying

$$\psi_i \wedge \varphi_i = 0.$$

Then

$$\psi_i = h_{ij}\varphi_j, \quad h_{ij} = h_{ji}.$$

Applying this to (ii) in (6), we conclude that on $\mathcal{F}(X)$

$$\omega^{\mu}_{i} = h^{\mu}_{ij}\omega^{j}, \quad h^{\mu}_{ij} = h^{\mu}_{ji}$$

for suitable functions h^{μ}_{ij} on $\mathcal{F}(X)$. The $r + nr$ relations

$$\begin{aligned} &\text{(i)} \quad \omega^{\mu} = 0 \\ &\text{(ii)} \quad \omega^{\mu}_{i} - h^{\mu}_{ij}\omega^{j} = 0 \end{aligned}$$

are the sought for equations defining $T\mathcal{F}(X) \subset T\mathcal{F}(\mathbf{E}^N)\big|_{\mathcal{F}(X)}$.

DEFINITION: The vector-valued quadratic differential form

$$II = h^{\mu}_{ij} e_{\mu} \otimes \omega^i \omega^j$$

is called the <u>2nd fundamental form</u> of $X \subset \mathbf{E}^{n+r}$.

It is well known that II is the basic extrinsic invariant of X in $\mathbf{E}^N$. Given $x \in X$ we may choose Euclidean coordinates $(y^1,\ldots,y^n,z^{n+1},\ldots,z^{n+r}) = (y^i;z^{\mu})$ centered at x such that X is given by

$$z^{\mu} = h^{\mu}_{ij} y^i y^j + (\ldots)$$

where $(\ldots)$ are higher order terms and

$$II(x) = h^{\mu}_{ij} \partial/\partial z^{\mu} \otimes dy^i dy^j.$$

Thus, II gives the quadratic approximation to X at each point.

Now let $\overline{X}$ be an abstract Riemannian manifold and $\mathcal{F}(\overline{X})$ the bundle of all orthonormal tangent frames $(y,\overline{e}_1,\ldots,\overline{e}_n) = (y;\overline{e}_i)$ where $y \in \overline{X}$ and $\overline{e}_i \in T_y(\overline{X})$. It is well known that there is on $\mathcal{F}(\overline{X})$ a coframing

$\bar{\omega}^i$, $\bar{\omega}^i_j$ $(i < j)$ satisfying

$$\text{(8)}\quad \begin{aligned} &\text{(i)} \quad \overline{ds}^2 = \sum_i (\bar{\omega}^i)^2 \\ &\text{(ii)} \quad d\bar{\omega}^i = -\bar{\omega}^i_j \wedge \bar{\omega}^j \end{aligned}$$

where $\overline{ds}^2$ is the pullback to $\mathcal{F}(\bar{X})$ of the 1st fundamental form of $\bar{X}$ and $(\bar{\omega}^i_j)$ is the connection matrix of the metric, or Levi-Civita connection. The curvature matrix $(\bar{\Omega}^i_j)$ is defined by the Cartan structure equation

$$d\bar{\omega}^i_j + \bar{\omega}^i_k \wedge \bar{\omega}^k_j = \bar{\Omega}^i_j .$$

It satisfies

$$\text{(9)}\quad \begin{aligned} &\bar{\Omega}^i_j + \bar{\Omega}^j_i = 0 \\ &\bar{\Omega}^i_j = \tfrac{1}{2} R_{ijk\ell} \bar{\omega}^k \wedge \bar{\omega}^\ell \end{aligned}$$

where $R_{ijk\ell}$ is the curvature tensor of $\bar{X}$, here viewed invariantly as a set of functions on $\mathcal{F}(\bar{X})$.

Suppose now that $\bar{X}$ admits an isometric embedding

$$\text{(10)}\quad \bar{X} \xrightarrow{x} X \subset \mathbf{E}^{n+r} .$$

We define

$$N \subset \mathcal{F}(\overline{X}) \times \mathcal{F}(\mathbf{E}^N) \tag{11}$$

to be the set of pairs of frames

$$((y,\overline{e}_i),(x,e_i,e_\mu)) \tag{12}$$

such that

$$\begin{aligned} x(y) &= x \\ x_*(\overline{e}_i) &= e_i \end{aligned} \tag{13}$$

where we have used x_* for the differential of the map x in (10). Then N is a principal $O(n) \times O(r)$ bundle over $\overline{X}$.

We want to prove that when restricted to N (i.e. when pulled back to N by the inclusion mapping)

$$\begin{aligned} &\text{(i)} && \omega^i - \overline{\omega}^i = 0 \\ &\text{(ii)} && \omega^\mu = 0 \\ &\text{(iii)} && \bigwedge_i \omega^i \bigwedge_{i<j} \omega^i_j \bigwedge_{\mu<\nu} \omega^\mu_\nu \neq 0 \end{aligned} \tag{14}$$

To prove (i) and (ii) it suffices to show that for

any local section u of the bundle $N \to \bar{X}$ we have

$$u^*(\omega^i - \bar{\omega}^i) = 0$$
$$u^*\omega^\mu = 0$$

(The forms $\bar{\omega}^i$ and ω^a are already zero on the fibres of N). Since u has the form (12), it is a pair of sections: $(y, \bar{e}_i)$ of $\mathcal{F}(\bar{X}) \to \bar{X}$ and (x, e_i, e_μ) of $\mathcal{F}(E^N) \to \bar{X}$.

By the first equation in (13)

$$dx = x_* \circ dy,$$

where dy is interpreted as the differential of the identity map of $\bar{X}$. Then

$$dy = ((y, \bar{e}_i)^* \bar{\omega}^i)\bar{e}_i, \quad \text{and}$$
$$dx = ((x, e_i, e_\mu)^* \omega^a) e_a;$$

and thus by the second equation in (13)

$$dx = ((y, \bar{e}_i)^* \bar{\omega}^i) e_i.$$

Comparing the last two expressions for dx, one verifies (i) and (ii).

We have not yet proved (iii), but geometrically it

is clear: it says that x is free to move infinitesimally in T_xX, and moreover that we are allowed to freely spin the tangent and normal frames.

Later on, we will see conversely that a submanifold $N \subset \mathcal{F}(\overline{X}) \times \mathcal{F}(E^N)$ of dimension $n(n + 1)/2 + r(r - 1)/2$ and satisfying (14) arises locally from an isometric embedding (10) by the above construction.

Summarizing

(15) <u>The local isometric embeddings (10) are in a one-to-one correspondence with submanifolds (11) satisfying (14)</u>.

We will study isometric embeddings as being solution manifolds to the differential system (14).

The most important feature of an isometric embedding are the Gauss equations relating the 2nd fundamental form (extrinsic) to the curvature (intrinsic). To derive them we set

$$\varphi^i_j = \omega^i_j - \overline{\omega}^i_j = -\varphi^j_i .$$

On a solution manifold N we have

$$\omega^i - \bar{\omega}^i = 0$$

$$\Rightarrow d(\omega^i - \bar{\omega}^i) = 0$$

$$\Rightarrow \omega^i_j \wedge \omega^j + \omega^i_\mu \wedge \omega^\mu - \bar{\omega}^i_j \wedge \bar{\omega}^j = 0$$

by the structure equations (3) and (8),

$$\Rightarrow \varphi^i_j \wedge \omega^j = 0$$

by (i), (ii) in (14),

$$\Rightarrow \varphi^i_j = a^i_{jk}\omega^k$$

by the Cartan lemma where

(16)
$$a^i_{jk} = a^i_{kj}$$
$$a^i_{jk} = -a^j_{ik}.$$

It is a well known elementary exercise that an a^i_{jk} with these symmetries must be zero. In conclusion, a consequence of the equations (14) is that

(17) $$\omega^i_j - \bar{\omega}^i_j = 0$$

on a solution manifold N . Using (3) and (14) the exterior derivatives of these equations give

$$\omega^{\mu}_{i} \wedge \omega^{\mu}_{j} = \bar{\Omega}^{i}_{j} \tag{18}$$

on N . When written out these are the famous <u>Gauss equations</u>

$$R_{ijk\ell} = h^{\mu}_{ik} h^{\mu}_{j\ell} - h^{\mu}_{i\ell} h^{\mu}_{jk}. \tag{19}$$

Setting

$$H_{ij} = h^{\mu}_{ij} e_{\mu} = H_{ji},$$

we may express the Riemannian curvature in terms of the 2×2 minors of the vector-valued symmetric matrix (H_{ij}):

$$R_{ijk\ell} = H_{ik} \cdot H_{j\ell} - H_{i\ell} \cdot H_{jk}. \tag{20}$$

Another useful notation is

$$R_{ijk\ell} = \gamma(H,H)_{ijk\ell} \tag{21}$$

where $\gamma(H,H)_{ijk\ell}$ is given by the right hand side of (20).

Note that even though the isometric embedding system (1.2) is 1st order, its most important geometric feature, the Guass equations, are 2nd order, and therefore we must differentiate the original system (i.e., prolong the system) in order to uncover the geometry.

REMARKS: There are two directions in which the receiving space can be generalized in this approach. Euclidean space $\mathbf{E}^N$ can be replaced by an arbitrary Riemannian manifold M. Then the bundle of orthonormal frames $\mathcal{F}(M)$ no longer has a group structure, except on its fibres, but it does possess a global coframing given by its canonical form (ω^a) and Levi-Civita connection form (ω^a_b). Of course now the structure equations will involve the curvature of M.

In the other direction $\mathbf{E}^N = E(N)/O(N)$ can be replaced by any homogeneous space $M = G/H$. Upon a choice of reference frame (o, e_a) of M, one has a bundle homomorphism $G \rightarrow \mathcal{F}(M)$ given by $u \rightarrow (u(o), u_* e_a)$. In general this homomorphism fails in interesting cases to be either surjective or injective, or both. Nevertheless, the Maurer-Cartan form of G gives a coframing on the image, the "adapted frames", and one can proceed in the same way to study the embedding problem for the appropriate abstract structures. Such an example is developed in Chapter 8. Compare also Chern [1], K. Yang [1], Markowitz [1], and also Jensen [1] for the related rigidity question.

CHAPTER 3

PFAFFIAN DIFFERENTIAL SYSTEMS

Although of considerable formal elegance and undeniable importance, the theory of differential systems is plagued by both a plethora of basic concepts and definitions, and at least some confusion concerning the basic notions of prolongation and involution. We will attempt to wade through a little of this without unduly aggravating the existing situation.

A Pfaffian differential system on a manifold M is given locally by

$$(1) \qquad \begin{aligned} &\text{(i)} \quad \theta^\alpha = 0 \qquad \alpha = 1,\ldots,s \\ &\text{(ii)} \quad \Omega = \omega^1\wedge\ldots\wedge\omega^p \neq 0. \end{aligned}$$

Here the θ^α, ω^ρ are everywhere linearly independent 1-forms. Clearly we should allow invertible substitutions

$$\tilde{\theta}^\alpha = A^\alpha_\beta\theta^\beta$$

$$\tilde{\omega}^\rho = B^\rho_\sigma\omega^\sigma + C^\rho_\alpha\theta^\alpha,$$

where throughout the remainder of these talks we shall use the additional ranges of indices

$$(2) \qquad \begin{aligned} &1 \le \alpha,\beta,\gamma \le s \\ &1 \le \rho,\sigma \le p. \end{aligned}$$

This motivates the following:

DEFINITION: A Pfaffian differential system (P.D.S.) on a manifold M is given by sub-bundles

$$I \subset J \subset T^*M.$$

Locally we will have

$$\begin{aligned} &\underline{I} = \operatorname{span}(\theta^\alpha) \\ &\cap \\ &\underline{J} = \operatorname{span}(\theta^\alpha;\omega^\rho). \end{aligned}$$

DEFINITION: An <u>integral manifold</u> of (I,J) is a p-dimensional submanifold $N \subset M$ such that

$$\text{(i)} \quad \theta^\alpha\big|_N = 0$$

$$\text{(ii)} \quad \Omega\big|_N \neq 0.$$

We may think of (i) in (1) as being a differential equation (this is clearly true in local coordinates), and then in particular an integral manifold is a solution. Condition (ii) is a transversality, or as we shall say, an independence condition on solution manifolds.

(3) EXAMPLE: Let $M = \mathcal{F}(E^{n+r})$ and give (I, J) in the form (1) by

$$\text{(i)} \quad \omega^{\mu} = 0 \qquad \mu = n+1, \ldots, n+r$$

$$\text{(ii)} \quad \bigwedge_{i} \omega^{i} \wedge \bigwedge_{i<j} \omega_{j}^{i} \bigwedge_{\mu<\nu} \omega_{\nu}^{\mu} \neq 0,$$

(here $p = n + n(n-1)/2 + r(r-1)/2$). It is not too difficult to see that integral manifolds are locally Darboux frame bundles $\mathcal{F}(X)$ of submanifolds $X^n \subset E^{n+r}$ (cf. the discussion of Cauchy characteristics at the end of this section).

(4) EXAMPLE: Let $M = \mathcal{F}(\bar{X}) \times \mathcal{F}(E^{n+r})$ and give (I, J) in the form (1) by

$$(5) \qquad \begin{aligned} &\text{(i)} \quad \omega^{i} - \bar{\omega}^{i} = 0 \\ &\text{(ii)} \quad \omega^{\mu} = 0 \\ &\text{(iii)} \quad \bigwedge_{i} \omega^{i} \wedge \bigwedge_{i<j} \omega_{j}^{i} \bigwedge_{\mu<\nu} \omega_{\nu}^{\mu} \neq 0 \end{aligned}$$

as in (2.14). A little argument, given near the end of this section, shows that the integral manifolds of this system are locally in one-to-one correspondence with isometric embeddings $\overline{X} \xrightarrow{x} X \subset \mathbf{E}^N$. We shall call (5) the naive isometric embedding system.

(6) EXAMPLE: Let Y be a manifold and $M = G_p(Y)$ the Grassmann bundle of p-planes in TY. We shall describe a canonical P.D.S. on M whose integral manifolds are in one-to-one correspondence with p-dimensional submanifolds of Y (keep in mind that all our discussion is local).

We may cover M by open sets of the following type: Let $(y^1,\ldots,y^p,z^1,\ldots,z^s) = (y^\rho;z^\alpha)$ be local coordinates on Y. Any p-plane $E \subset TY$ on which the restrictions of the dy^ρ remain linearly independent has equations

$$dz^\alpha - \ell^\alpha_\rho dy^\rho = 0,$$

and we may take

$$(y^\rho;z^\alpha;\ell^\alpha_\rho)$$

as local coordinates on $M = G_p(Y)$. The P.D.S. given in

the form (1) by

$$\begin{aligned} \theta^\alpha &= dz^\alpha - \ell^\alpha_\rho dy^\rho = 0 \\ \Omega &= dy^1 \wedge \ldots \wedge dy^p \neq 0 \end{aligned} \tag{7}$$

has intrinsic meaning and gives the <u>canonical system</u> (I, J) on $G_p(Y)$. (Note: Points in $G_p(Y)$ are $x = (y, E)$ where $y \in Y$ and $E \subset T_y Y$ is a p-plane. If $\pi : G_p(Y) \to Y$ is the projection, then $I_x \subset J_x \subset T^*_x M$ is given by

$$\begin{aligned} I_x &= \pi^*(E^\perp) \\ &\cap \\ J_x &= \pi^*(T^*_y Y) \end{aligned}$$

where $E^\perp \subset T^*_y Y$ is the annihilator of E.)

An integral manifold N of the canonical system is locally given by

$$(y^\rho) \to (y^\rho; z^\alpha(y); \ell^\alpha_\rho(y)).$$

This is because $dy^1 \wedge \ldots \wedge dy^p |_N \neq 0$ so that $y^1, \ldots, y^p$ gives a local coordinate system on N. The condition

$$0 = \theta^\alpha |_N = \left(\frac{\partial z^\alpha}{\partial y^\rho}(y) - \ell^\alpha_\rho(y) \right) dy^\rho$$

gives

$$\ell^{\alpha}_{\rho}(y) = \frac{\partial z^{\alpha}}{\partial y^{\rho}}(y).$$

In conclusion, if for a p-dimensional submanifold $Z^p \subset Y$ we define its <u>1st prolongation</u> (sometimes denoted by $Z^{(1)}$), to be the submanifold $N \subset G_p(Y)$ given by

$$N = \{(y, T_y Z) : y \in Z\},$$

then:

<u>the integral manifolds of the canonical system on $G_p(Y)$ are locally just the 1st prolongations of p-dimensional submanifolds</u> $Z \subset Y$.

(8) EXAMPLE: Finally, if (I', J') is a P.D.S. on a manifold M', then under suitable (fairly obvious) transversality conditions we may restrict this system to a submanifold $M \subset M'$ to obtain a P.D.S. (I, J) there.

For instance, suppose that $Y = \mathbb{C}^m$, and $M \subset G_{2q}(\mathbb{C}^m)$ consists of the <u>complex</u> q-planes among the real 2q-planes in $T\mathbb{C}^m$. The integral manifolds of the restriction to M of the canonical system on $G_{2q}(\mathbb{C}^m)$ are then in one-to-one correspondence with the complex q-dimensional submanifolds of $\mathbb{C}^m$.

A general submanifold $M \subset G_p(Y)$ is locally given by equations

$$F^\lambda(y^\rho; z^\alpha; \ell^\alpha_\rho) = 0.$$

Integral manifolds of the restriction to M of the canonical system therefore are locally just solutions to the P.D.E. system

$$F^\lambda\left[y^\rho; z^\alpha; \frac{\partial z^\alpha}{\partial y^\rho}(y)\right] = 0.$$

In fact, using jet bundles there is a canonical way of writing any P.D.E. system as a P.D.S. (cf. Bryant et al. [1] and examples (24) and (4.13) below).

We now return to the general discussion of a P.D.S. (I,J) on M. Denote by $\Lambda^q M$ the C^∞ q-forms on M and by $\underline{I} \subset \Lambda^1 M$ the C^∞ sections of I. From the basic observations

$$\theta\big|_N = 0 \Rightarrow d\theta\big|_N = 0$$

(9)

$$\theta\big|_N = 0 \Rightarrow \varphi\wedge\theta\big|_N = 0 \quad \text{for all} \quad \varphi \in \Lambda^* M = \oplus\Lambda^q M$$

we are led to consider the differential ideal $\mathcal{I} \subset \Lambda^* M$ generated by $\underline{I}$. Recall that a differential ideal is a homogeneous ideal $\mathcal{I} = \oplus \mathcal{I}^q \subset \Lambda^* M$ that is closed under d. In our case, $\mathcal{I}$ is algebraically generated by the 1-forms $\theta \in \underline{I}$ and their exterior derivatives. We may summarize (9) by saying that:

(10) If N is an integral manifold of (I,J), then

$$\varphi\big|_N = 0 \quad \text{for all} \quad \varphi \in \mathcal{I}.$$

This is equivalent to

$$\begin{aligned} \theta\big|_N &= 0 \quad \text{and} \\ d\theta\big|_N &= 0 \end{aligned}$$

for all $\theta \in \underline{I}$.

The theory of exterior differential systems seeks to build up integral manifolds by "integrating" their potential tangent spaces. Here the crucial concept is given by the

DEFINITIONS i) An <u>integral element</u> of a differential ideal $\mathcal{I}$ is a subspace $E \subset T_x M$ satisfying

(11) $$\varphi(x)\big|_E = 0$$

for all $\varphi \in \mathcal{I}$.

ii) Given a P.D.S. (I,J), the set of p-dimensional integral elements $E \subset T_xM$ of $\mathcal{I}$ satisfying

$$\Omega(x)\big|_E \neq 0$$

will be denoted by $V(I,J)$, and will be called simply the integral elements of (I,J).

In a natural way

$$V(I,J) \subset G_p(M);$$

for simplicity of exposition we will assume that $V(I,J)$ is a submanifold (in practice one generally restricts to smooth open subsets of $V(I,J)$). An integral manifold of (I,J) is now given by a submanifold $N \subset M$ whose 1st prolongation $N^{(1)}$ lies in $V(I,J) \subset G_p(M)$.

Three of the most important ingredients in the theory of differential systems are i) the Cartan-Kähler theorm; ii) involution; and iii) prolongation and the Cartan-Kuranishi theorem. We will briefly explain each, referring to Bryant et al. [1] for more precise definitions and proofs.

The Cartan-Kähler theorem deals with a differential ideal $\mathcal{I} \subset \Lambda^* M$ (no independence condition). Denote by $V_q(\mathcal{I}) \subset G_q(M)$ the set of q-dimensional integral elements of $\mathcal{I}$. Points in $G_q(M)$ will be written as (x, E^q) where $E^q \subset T_x M$. Given $(x, E^q) \in V_q(\mathcal{I})$ with basis $e_1, \ldots, e_q$ for E^q, its <u>polar equations</u> are the linear equations for the subspace of all $v \in T_x M$ which satisfy

$$(12) \qquad \langle \varphi(x), e_1 \wedge \ldots \wedge e_q \wedge v \rangle = 0, \quad \varphi \in \mathcal{I}^{q+1}.$$

In other words, the polar equations of (x, E^q) are the elements of the subspace of $T_x^* M$ spanned by

$$\{\iota(e_1 \wedge \ldots \wedge e_q)\varphi : \varphi \in \mathcal{I}^{q+1}\}.$$

The solutions v which do not lie in E^q are precisely the vectors by which E^q can be enlarged to a (q+1)-dimensional integral element.

DEFINITION: The integral element (x, E^q) is <u>K-regular</u> if i) (x, E^q) is a smooth point of $V_q(\mathcal{I})$, and ii) the rank of the polar equations (12) is constant near (x, E^q).

Here, the defining equations of $V_q(\mathcal{I})$ are (11), and to be a smooth point means that the usual Jacobian criterion is satisfied for these defining equations.

A special case of the Cartan-Kähler theorem is the following result.

(13) Let $\mathcal{I}$ be a real analytic differential ideal on a real analytic manifold, and suppose that $(x,E) \in V_p(\mathcal{I})$ satisfies the condition:

(14) There exists a flag

$$(0) = E^0 \subset E^1 \subset \ldots \subset E^{p-1} \subset E^p = E, \quad \dim E^q = q,$$

consisting of K-regular integral elements (x,E^q) for $q = 0,1,\ldots,p-1$.

Then there exists a local real analytic integral manifold N of $\mathcal{I}$ passing through x and with $T_xN = E$.

An integral element satisfying the condition (14) is said to <u>admit a regular flag</u>.

Intuitively, N is built up from a nested sequence

of integral manifolds

$$N^0 \subset N^1 \subset \dots \subset N^{p-1} \subset N^p$$

where $x \in N^q$ and $T_x N^q = E^q$; moreover, N^{q+1} is constructed from N^q by solving a determined P.D.E. system of Cauchy-Kowaleska type (whence the real analytic assumption). The general Cartan-Kähler theorem also tells the degree of freedom present in constructing N^{q+1} from N^q; in this sense it tells us "how many" local integral manifolds of $\mathcal{I}$ there are.

The concept of involution is probably the most difficult one to understand in the theory. The definition is, however, easy to give:

DEFINITION: The Pfaffian differential system (I,J) is <u>involutive</u> if a general integral element $(x,E) \in V(I,J)$ admits a regular flag.

Here, "general" means in an open dense set of $V(I,J)$. Note that the definition of involutive requires an independence condition.

By the Cartan-Kähler theorem, an involutive, real analytic P.D.S. has local integral manifolds.

We shall now try to at least partially illuminate

the intuitive meaning of involutivity. Suppose that the P.D.S. is given locally by

$$(15)\quad \begin{aligned} &\text{(i)} \quad \theta^{\alpha} = 0 \\ &\text{(ii)} \quad \Omega = \omega^1 \wedge \ldots \wedge \omega^p \neq 0. \end{aligned}$$

Under (i) it is understood that we adjoin all exterior equations arising from the differential ideal $\mathcal{I}$ generated by the θ^{α}'s. Roughly speaking, <u>that (I,J) is involutive means that the equations (15) do not imply any additional equations</u>.

In practice there are two ways that (15) may imply additional equations. The first is given by the following general

(16) CONSTRUCTION: Let $M' = \{x \in M$: there exists an integral element in $V(I,J)$ over $x\}$ be the image of

$$V(I,J) \to M;$$

(for simplicity we assume M' is a manifold). If $M = M'$ then we do nothing. However, if $M' \neq M$ then we note that any integral manifold of (I,J) must lie in M', and so we set

$$V'(I,J) = \{(x,E) \in V(I,J): E \subset T_x M'\}.$$

Now consider the projection

$$V'(I,J) \to M'$$

with image M''. If $M'' = M'$ we stop; otherwise we continue as before. Eventually we arrive either at the empty set (in which case (I,J) has no integrals), or else at M_1 with $V_1(I,J) \to M_1$ being surjective and with all $(x,E) \in V_1(I,J)$ satisfying $E \subset T_x M_1$. Clearly we must add the equations of $M_1 \subset M$ to (I,J). In particular, if $M' \neq M$ then the system is not involutive.

The simplest example of this construction is when $\theta^1,\dots,\theta^s,\omega^1,\dots,\omega^p$ are a coframe for M. The condition that there be an integral element through each point $x \in M$ is the

Frobenius condition

$$(17) \qquad d\theta^\alpha \equiv 0 \bmod\{\theta\},$$

where $\{\theta\} \subset \Lambda^* M$ is the algebraic (not differential) ideal generated by $\underline{I}$. In this case, (17) is equivalent to involutivity.

Now suppose that over each point $x \in M$ there is an integral element. Then it may still happen that the equations (15) imply additional exterior equations.

(18) EXAMPLE: For the naive isometric embedding system (5), the argument just above (2.16) shows that the original system implies the additional equations (2.17 and 2.18). Thus (5) is not involutive (below we shall see this in another way).

Finally, we want to define prolongation.

DEFINTIONS: i) Let (I,J) be a Pfaffian differential system on a manifold M. Assuming that $V(I,J) \to M$ is surjective, we set $M^{(1)} = V(I,J)$ and define the <u>1st prolongation</u> $(I^{(1)},J^{(1)})$ to be the restriction to $M^{(1)}$ of the canonical P.D.S. on $G_p(M)$. In general, we set $M^{(1)} = V_1(I,J)$ as in the construction (16) and then, as above, $(I^{(1)},J^{(1)})$ is the restriction to $M^{(1)}$ of the canonical system.

ii) The <u>q^{th} prolongation</u> $(I^{(q)},J^{(q)})$ on $M^{(q)}$ is defined inductively to be the 1st prolongation of $(I^{(q-1)},J^{(q-1)})$ on $M^{(q-1)}$.

The meaning of this definition becomes more transparent if one writes it out in more detail.

Suppose that the P.D.S. (I,J) is given locally by

$$\text{(i)}\quad \theta^{\alpha} = 0$$
$$\text{(ii)}\quad \Omega = \omega^{1}\wedge\ldots\wedge\omega^{p} \neq 0,$$

and let $\pi^{1},\ldots,\pi^{r} = \{\pi^{\epsilon}\}$ be a set of 1-forms that completes $\{\theta^{\alpha};\omega^{\rho}\}$ to a local coframe. Then we have

$$d\theta^{\alpha} \equiv \frac{1}{2}\, a^{\alpha}_{\epsilon\delta}\pi^{\epsilon}\wedge\pi^{\delta} + b^{\alpha}_{\epsilon\rho}\pi^{\epsilon}\wedge\omega^{\rho} + \frac{1}{2}\, c^{\alpha}_{\rho\sigma}\omega^{\rho}\wedge\omega^{\sigma} \quad \mathrm{mod}\{\theta\},$$

where $a^{\alpha}_{\epsilon\delta} = -a^{\alpha}_{\delta\epsilon}$, $c^{\alpha}_{\rho\sigma} = -c^{\alpha}_{\sigma\rho}$, and $\{\theta\}$ is the algebraic ideal generated by the θ^{α}.

Any element $(x,E) \in G_p(M)$ for which $\Omega|_E \neq 0$ has equations

$$\theta^{\alpha} = m^{\alpha}_{\rho}\omega^{\rho}, \quad \pi^{\epsilon} = \ell^{\epsilon}_{\rho}\omega^{\rho}.$$

Thus $\{m^{\alpha}_{\rho};\ell^{\epsilon}_{\rho}\}$ are local fibre coordinates in $G_p(M)$. The canonical P.D.S. on $G_p(M)$ is given locally by

$$\text{(i)}\quad \theta^{\alpha} - m^{\alpha}_{\rho}\omega^{\rho} = 0, \quad \pi^{\epsilon} - \ell^{\epsilon}_{\rho}\omega^{\rho} = 0$$

$$\text{(ii)}\quad \Omega \neq 0,$$

where, as usual, we make no distinction between forms on M and their lift to $G_p(M)$ by the projection mapping.

Now $(x,E) \in V(I,J)$ if and only if

$$\theta^\alpha\Big|_E = 0 = d\theta^\alpha\Big|_E \quad\text{and}\quad \Omega\Big|_E \neq 0;$$

which means

$$(*)\qquad \begin{aligned} &m^\alpha_\rho = 0, \quad\text{and}\\ &a^\alpha_{\epsilon\delta}(\ell^\epsilon_\rho\ell^\delta_\sigma - \ell^\epsilon_\sigma\ell^\delta_\rho) + (b^\alpha_{\epsilon\sigma}\ell^\epsilon_\rho - b^\alpha_{\epsilon\rho}\ell^\epsilon_\sigma) + c^\alpha_{\rho\sigma} = 0.\end{aligned}$$

Thus $(*)$ are the local equations of $V(I,J) \subset G_p(M)$. If we assume that $(*)$ has solutions for each $x \in M$, so that $M^{(1)} = V(I,J)$, then the 1st prolongation $(I^{(1)},J^{(1)})$ of (I,J) is given locally on $M^{(1)}$ by

$$\begin{aligned} &\text{i)}\quad \theta^\alpha = 0,\quad \pi^\epsilon - \ell^\epsilon_\rho\omega^\rho = 0\\ &\text{ii)}\quad \Omega \neq 0.\end{aligned}$$

Put another way,

$$\underline{I}^{(1)} = \operatorname{span}(\theta^\alpha,\ \pi^\epsilon - \ell^\epsilon_\rho\omega^\rho)$$

$$\underline{J}^{(1)} = \operatorname{span}(\underline{I}^{(1)},\omega^\rho) = \operatorname{span}(\underline{J},\pi^\epsilon) = \text{lift of } T^*M.$$

Observe that $d\underline{I}^{(1)} \subset \{J^{(1)}\}$ = the algebraic ideal generated by $J^{(1)}$. Thus the 1st prolongation of any

P.D.S. (I,J) is quasilinear, (see §4 where this important concept is introduced).

From example (6) and the construction (16) it is clear that:

(19) <u>The integral manifolds of a P.D.S. and of its 1st prolongation are in a one-to-one correspondence</u>.

Other general properties of prolongations will be given below.

A basic result concerning prolongations is the

(20) THEOREM (Cartan-Kuranishi): Given a Pfaffian differential system (I,J), there is a q_0 such that prolongations $(I^{(q)},J^{(q)})$ are involutive for $q \geq q_0$.

Actually, some mild regularity hypotheses are required, but we have no need to specify these.

For us the importance of prolongation resides mainly in the following observation: For the isometric embedding problem, when $N < N(n) = n(n+1)/2$ the system (1.2) is overdetermined. Hence there must be integrability conditions, which for geometric reasons must be expressed as relations on $R(x)$, $\nabla R(x)$, $\nabla^2 R(x), \dots,$ in order that even a formal embedding should exist about each point (formal means formal power series). Prolongation gives a method for finding these conditions. However, as we shall see, it is better not to prolong directly but to go via the characteristic variety. Passing to the characteristic variety has, in a certain sense, the effect of handling all the prolongations at once, so that one is not faced with the practically impossible task of differentiating the isometric embedding equations (i.e., prolonging) a large number of times.

APPENDIX TO §3: Cauchy characteristics (cf. Byrant et al. [1]). Let M be a manifold, $\mathcal{I} \subset \Lambda^{*}M$ a differential ideal, and $V(M)$ the Lie algebra of C^{∞} vector fields on M. We define

$$A(\mathcal{I}) \subset V(M)$$

by

$$A(\mathcal{I}) = \{v \in V(M) : \iota(v)\mathcal{I} \subseteqq \mathcal{I}\}.$$

Since $d\mathcal{I} \subseteqq \mathcal{I}$, from the Cartan formula

$$\mathcal{L}_{v}(\varphi) = d\iota(v)\varphi + \iota(v)d\varphi$$

for the Lie derivative $\mathcal{L}_{v}\varphi$ of a form φ with respect to a vector field v, we see that intuitively

$$A(\mathcal{I}) = \{v \in V(M) : \mathcal{L}_v(\mathcal{I}) \subseteqq \mathcal{I}\}.$$

It may readily be seen that $A(\mathcal{I})$ is a sub-Lie algebra. Set

$$A_x = \{v(x) : v \in A(\mathcal{I})\} \subset T_xM$$

and assume that $\dim A_x$ is constant. Then

$$A = \bigcup_x A_x \subset TM$$

is an integrable distribution, called the <u>Cauchy characteristic space</u> associated to $\mathcal{I}$. Assume that the foliation given by A is a fibration (this is true locally), say

$$f : M \to M'.$$

A basic fact is that there is a differential ideal $\mathcal{I}' \subset \Lambda^* M'$ such that $f^*\mathcal{I}'$ generates $\mathcal{I}$ algebraically. In coordinates, $M \to M'$ will locally be given by

$$(y^1,\ldots,y^k,z^1,\ldots,z^\ell) \to (y^1,\ldots,y^k),$$

and $\mathcal{I}$ will be generated by C^∞ forms in the y-variables. It follows that

(21) If $N' \subset M'$ is an integral manifold of $\mathcal{I}'$, then $N = f^{-1}(N') \subset M$ is an integral manifold of $\mathcal{I}$.

In example (4) we let $\partial/\partial\omega^i$, $\partial/\partial\omega^i_j$, $\partial/\partial\overline{\omega}^i, \ldots$ be the tangent frame on $\mathcal{F}(\overline{X}) \times \mathcal{F}(E^N)$ dual to ω^i, ω^i_j, $\overline{\omega}^i, \ldots$. Then it is a nice exercise, using the stucture equations (5.1) below, to show that:

(22) The Cauchy characteristic space of (5) is spanned by the vector fields $\partial/\partial\omega^i_j$, $\partial/\partial\omega^\mu_\nu$.

Geometrically, the leaves of the Cauchy characteristic foliation correspond to spinning the tangent and normal frames. (See (5.31) for a discussion of the Cauchy characteristics of (3)).

Now suppose that for every p-dimensional integral element $E \subset T_xM$ of (I, J) it is the case that

(23) $$A_x \subset E.$$

Then there is a converse to (21): every integral manifold N is (locally) $f^{-1}(N')$ for an integral manifold N' of $\mathcal{I}'$. This is clear from (23) and the integrability of A. When we compute the integral elements of (5) in §5 below it will be seen, using (22), that (23) is valid. Thus, the integral manifolds N of

(5) on $\mathcal{F}(\overline{X}) \times \mathcal{F}(\mathbf{E}^N)$ consist of manifolds of frames invariant under $O(n) \times O(r)$, where $O(n)$ acts on the frame $\{\overline{e}_i\}$ and the frame $\{e_i\}$ by the same matrix (these frames spin at the same rate). Thus we have a diagram

$$\begin{array}{cccc} & N & \subset & \mathcal{F}(\overline{X}) \times \mathcal{F}(\mathbf{E}^N) \\ f & \downarrow & & \downarrow \\ & N' & \subset & \overline{X} \times \mathbf{E}^N \end{array}$$

where f is the Cauchy characteristic fibration. It is then more or less clear that $N' \subset \overline{X} \times \mathbf{E}^N$ is the graph of an isometric embedding.

We illustrate these ideas with the familiar example of a first order P.D.E. (Compare John [1]).

(24) EXAMPLE: Consider the P.D.E.

$$F(x^i, u, \frac{\partial u}{\partial x^i}) = 0, \quad i,j = 1,\ldots,n.$$

Let

$$M = \{F(x^i, z, p_i) = 0\} \subseteqq \mathbb{R}^{2n+1}.$$

There is no substantial loss in generality in assuming that $F_{p_n} \neq 0$ on M. Then M is a smooth 2n-manifold.

We convert the P.D.E. to the P.D.S. on M given

locally by (all forms are restricted to M)

(25)
$$\text{i)}\quad \theta = dz - p_i dx^i$$
$$\text{ii)}\quad dx^1 \wedge \ldots \wedge dx^n \neq 0.$$

Then

(26)
$$d\theta = -dp_i \wedge dx^i$$

and the symbol relations (see (4.11)) are

(27)
$$0 = dF\big|_M = F_{x_i} dx^i + F_z dz + F_{p_i} dp_i$$

The differential ideal $\mathcal{I}$ generated by θ is the algebraic ideal generated by θ and $d\theta$. To find the Cauchy characteristics of (25), we want to find

$$\{v \in V(M) : \iota(v)\mathcal{I} \subseteqq \mathcal{I}\}.$$

But

$$\iota(v)\mathcal{I} \subseteqq \mathcal{I} \Leftrightarrow \begin{cases} \theta(v) = \theta & \text{and} \\ \iota(v)d\theta \in \text{span}(\theta). \end{cases}$$

Put

$$v = v^i \frac{\partial}{\partial x^i} + v_o \frac{\partial}{\partial z} + v_i \frac{\partial}{\partial p_i}.$$

Then $\theta(v) = 0$ gives

(28) $$v_0 = p_i v^i,$$

and $\iota(v)d\theta \in \operatorname{span}(\theta)$ gives

(29) $$\iota(v)d\theta = -v_i dx^i + v^i dp_i = \lambda\theta, \quad \text{some } \lambda \in \mathbb{R}.$$

If we solve for dp_n in (27), and substitute this into (29) we obtain

(30) $$\left(v_i - \lambda p_i + \frac{v_n F_{x^i}}{F_{p_n}}\right)dx^i + \left(\lambda + \frac{v_n F_z}{F_{p_n}}\right)dz + \sum_1^{n-1}\left(\frac{v_n F_{p_i}}{F_{p_n}} - v^i\right)dp_i = 0.$$

As $dx^1,\ldots,dx^n,dz,dp_1,\ldots,dp_{n-1}$ are linearly independent on M, (30) gives

(31) $$v^i = \frac{v_n F_{p_i}}{F_{p_n}}, \quad v_i = -\frac{v_n}{F_{p_n}}(F_z p_i + F_{x^i}).$$

Combining (28) and (31) we see that the Cauchy characteristic space associated to $\mathcal{I}$ is spanned at each point of M by the vector field

(32) $$v = F_{p_i}\frac{\partial}{\partial x^i} + p_i F_{p_i}\frac{\partial}{\partial z} + (F_{x^i} + F_z p_i)\frac{\partial}{\partial p_i}.$$

To relate this to the classical notions, we begin with the case of a quasi-linear P.D.E.

$$F(x^i, u, \frac{\partial u}{\partial x^j}) = a^i(x,u) \frac{\partial u}{\partial x^i} - c(x,u),$$

where we write $x = (x^1, \ldots, x^n)$. In this case

$$F_{p_i} = a^i(x,z) \quad \text{and} \quad a^i p_i = c(x,u) \quad \text{on} \quad M.$$

Hence v of (32) becomes

$$(33) \quad v = a^i(x,z) \frac{\partial}{\partial x^i} + c(x,z) \frac{\partial}{\partial z} + (F_{x^i} + F_z p_i) \frac{\partial}{\partial p_i},$$

and thus v projects onto the (x,z)-subspace:

$$\tilde{v} = \mathrm{Proj}_{\mathbb{R}^{n+1}} v = a^i(x,z) \frac{\partial}{\partial x^i} + c(x,z) \frac{\partial}{\partial z}.$$

Classically $\tilde{v}$ is called the Cauchy characteristic vector field of the quasi-linear P.D.E. If $z = u(x)$ is a solution to the P.D.E., then it is easily verified that $\tilde{v}(x,u(x))$ is tangent to the graph of $z = u(x)$ for every x. It follows that this graph is a union of integral curves of $\tilde{v}$. Thus the solution of the Cauchy problem for the quasi-linear P.D.E. is reduced to finding the integral curves of $\tilde{v}$ with specified initial conditions, and this involves only ordinary differential equations.

For the general P.D.E. v does not project onto (x,z)-space, as F_{p_i} and $p_i F_{p_i}$ will depend on the p_j in general. Now we must interpret (x,z,p)-space as the Grassmann bundle of tangent n-planes to $\mathbb{R}^{n+1}$. A solution $z = u(x)$ to the P.D.E. defines an integral submanifold to the P.D.S. (25):

$$(34) \qquad f(x) = (x^i, u(x), \frac{\partial u}{\partial x^j}),$$

which is the first prolongation of the graph of $z = u(x)$.

It is an elementary exercise to show that v is tangent to the submanifold (34); i.e., that

$$v(x^i, u(x), \frac{\partial u}{\partial x^j}) \in f_* T_x \mathbb{R}^n, \quad \forall x.$$

It follows then that the first prolongation of the graph of the solution $z = u(x)$ of the P.D.E. is a union of integral curves of v. Classically the integral curves of v are called characteristic strips.

CHAPTER 4

QUASI-LINEAR PFAFFIAN DIFFERENTIAL SYSTEMS

We consider a P.D.S. (I,J) on a manifold M. Among these systems there is a remarkable class given by the following:

DEFINITION: The system is said to be quasi-linear in case

$$d\underline{I} \subset \{\underline{J}\},$$

where $\{\underline{J}\} \subset \Lambda^{*}M$ is the algebraic ideal generated by $\underline{J}$.

To explain what this really means, we assume that (I,J) is given as before by

$$(1) \qquad \begin{aligned} &(i) \quad \theta^{\alpha} = 0 \\ &(ii) \quad \Omega = \omega^{1}\wedge\ldots\wedge\omega^{p} \neq 0, \end{aligned}$$

and we let $\pi^{1},\ldots,\pi^{r} = \{\pi^{\epsilon}\}$ be a set of 1-forms that completes $\{\theta^{\alpha};\omega^{\rho}\}$ to a local coframe. Then we always have relations, called the <u>structure equations</u> of the

P.D.S.,

$$(2)\quad d\theta^{\alpha} \equiv \frac{1}{2} b^{\alpha}_{\epsilon\delta}\pi^{\epsilon}\wedge\pi^{\delta} + a^{\alpha}_{\epsilon\rho}\pi^{\epsilon}\wedge\omega^{\rho} + \frac{1}{2} c^{\alpha}_{\rho\sigma}\omega^{\rho}\wedge\omega^{\sigma} \quad \mathrm{mod}\{\theta\}$$

where

$$b^{\alpha}_{\epsilon\delta} + b^{\alpha}_{\delta\epsilon} = 0 = c^{\alpha}_{\rho\sigma} + c^{\alpha}_{\sigma\rho}.$$

and $\{\theta\}$ is the algebraic ideal generated by the θ^{α}. The condition

$$(3)\qquad b^{\alpha}_{\epsilon\delta} = 0$$

is invariant under changes of coframe preserving the filtration $I \subset J \subset T^{*}M$; it is clearly equivalent to the condition that (I,J) be quasi-linear.

To gain a little understanding of the importance of this condition, let us look for the equations that define integral elements. We work over a fixed point $x \in M$ and will generally omit reference to it. Any p-plane $E \subset T_{x}M$ satisfying (i) and (ii) in (1) is given by linear equations (i) and

$$\pi^{\epsilon} - \ell^{\epsilon}_{\rho}\omega^{\rho} = 0.$$

The condition that this plane be an integral element is

$$d\theta^{\alpha}\Big|_{E} = 0;$$

by (2) this is

$$b^{\alpha}_{\epsilon\delta}(\ell^{\epsilon}_{\rho}\ell^{\delta}_{\sigma} - \ell^{\epsilon}_{\sigma}\ell^{\delta}_{\rho}) + (a^{\alpha}_{\epsilon\sigma}\ell^{\epsilon}_{\rho} - a^{\alpha}_{\epsilon\rho}\ell^{\epsilon}_{\sigma}) + c^{\alpha}_{\rho\sigma} = 0.$$

In the quasi-linear case these quadratic equations reduce to the linear equations

(5) $$a^{\alpha}_{\epsilon\sigma}\ell^{\epsilon}_{\rho} - a^{\alpha}_{\epsilon\rho}\ell^{\epsilon}_{\sigma} + c^{\alpha}_{\rho\sigma} = 0.$$

It goes without saying that treating linear equations is much simpler than treating quadratic ones.

Suppose now that (I,J) is quasi-linear and write (2) as

(6) $$d\theta^{\alpha} \equiv a^{\alpha}_{\epsilon\rho}\pi^{\epsilon}\wedge\omega^{\rho} + \frac{1}{2}\, c^{\alpha}_{\rho\sigma}\omega^{\rho}\wedge\omega^{\sigma} \quad \mathrm{mod}\{\theta\}$$

where

$$c^{\alpha}_{\rho\sigma} + c^{\alpha}_{\sigma\rho} = 0.$$

Consider the vector-valued 2-form

$$\tau = \left\{ \frac{1}{2}\, c^{\alpha}_{\rho\sigma}\omega^{\rho}\wedge\omega^{\sigma} \right\} \quad \mathrm{mod}\{\theta\}.$$

Under a change of coframe

$$\tilde{\theta}^\alpha = \theta^\alpha$$
$$\tilde{\omega}^\rho = \omega^\rho$$
$$\tilde{\pi}^\epsilon = \pi^\epsilon - \ell^\epsilon_\rho \omega^\rho$$

we have

$$\tilde{\tau} = \tau - \left\{ \tfrac{1}{2} (a^\alpha_{\epsilon\rho} \ell^\epsilon_\sigma - a^\alpha_{\epsilon\sigma} \ell^\epsilon_\rho) \omega^\rho \wedge \omega^\sigma \right\}; \tag{7}$$

we denote by $[\tau]$ the equivalence class of τ under substitutions (7).

DEFINITION: $[\tau]$ is called the <u>torsion</u> of (I,J).

Comparing (5) and (7) we see that:

(8) The vanishing of the torsion at $x \in M$ is equivalent to the existence of an integral element $E \subset T_xM$.

Clearly, the construction (3.16) consists in successively setting the torsion equal to zero (or, as we sometimes say, <u>annihilating the torsion</u>).

Before discussing some examples, we want to define

the other basic invariant of a quasi-linear system. For this we set

$$\pi^\alpha_\rho = -(a^\alpha_{\epsilon\rho}\pi^\epsilon - \frac{1}{2} c^\alpha_{\rho\sigma}\omega^\sigma)$$

and write (6) as

(9) $$d\theta^\alpha \equiv -\pi^\alpha_\rho \wedge \omega^\rho \quad \mathrm{mod}\{\theta\}.$$

The 1-forms π^α_ρ are not linearly independent in general. At a point $x \in M$, let $V^* = \mathrm{span}(\theta^\alpha)$, and let $\frac{\partial}{\partial\theta^\alpha}$ denote the dual basis of its dual space V. Furthermore, let $W^* = \mathrm{span}(\omega^\rho)$, and let $\frac{\partial}{\partial\omega^\rho}$ denote the dual basis of its dual space W. We define a subspace $\mathcal{A}$ of $V \otimes W^*$ by

$$\mathcal{A} = \mathrm{span}(a^\alpha_{\epsilon\rho} \frac{\partial}{\partial\theta^\alpha} \otimes \omega^\rho : \epsilon = 1,\dots,r).$$

It is straightforward to check that, under a change of coframe adapted to the filtration $I \subset J \subset T^*M$, the subspace $\mathcal{A}$ is invariantly defined. The annihilator $\mathcal{A}^\perp$ of $\mathcal{A}$ is a subspace of $V^* \otimes W$.

If we choose a basis r^λ of $\mathcal{A}^\perp$, then

$$r^\lambda = r^{\lambda\rho}_\alpha \theta^\alpha \otimes \frac{\partial}{\partial\omega^\rho},$$

and these coefficients satisfy

$$r^{\lambda\rho}_{\alpha} a^{\alpha}_{\epsilon\rho} = 0, \quad \forall \epsilon. \tag{10}$$

These coefficients will depend smoothly on x, and thus we have

$$r^{\lambda\rho}_{\alpha} \pi^{\alpha}_{\rho} \equiv 0 \quad \mathrm{mod}\{\theta, \omega\} \tag{11}$$

where, as always, $\{\theta, \omega\}$ is the algebraic ideal generated by the 1-forms $\theta^{\alpha}, \omega^{\rho}$.

DEFINITIONS: i) The relations (11) are called the <u>symbol relations</u> of (I, J);

ii) the matrix

$$\pi = \begin{Vmatrix} \pi^1_1 & \cdots & \pi^1_p \\ \vdots & & \vdots \\ \pi^s_1 & \cdots & \pi^s_p \end{Vmatrix}$$

is called the <u>tableau matrix</u> of (I, J).

(13) EXAMPLE: We consider a 1st order P.D.E. system

$$F^{\lambda}\left(y^{\rho}, z^{\alpha}, \frac{\partial z^{\alpha}(y)}{\partial y^{\rho}}\right) = 0. \tag{14}$$

In the space of variables $(y^\rho, z^\alpha, p^\alpha_\rho)$ let M (assumed to be a smooth manifold) be defined by

$$F^\lambda(y^\rho, z^\alpha, p^\alpha_\rho) = 0,$$

and set

$$\theta^\alpha = dz^\alpha - p^\alpha_\rho dy^\rho |_M$$
$$\omega^\rho = dy^\rho |_M$$
$$\pi^\alpha_\rho = dp^\alpha_\rho |_M .$$

Then the P.D.S. on M given by

$$\theta^\alpha = 0$$
$$\Omega = \omega^1 \wedge \ldots \wedge \omega^p \neq 0$$

has as integral manifolds locally the solutions to (14) (compare with example (3.6) and (3.24)). As clearly

$$d\theta^\alpha = -\pi^\alpha_\rho \wedge \omega^\rho ,$$

this system is quasi-linear and comes naturaly in the form (9); to obtain the form (6), we simple let π^ϵ be a maximal linearly independent set from among the π^α_ρ and write

$$\pi^\alpha_\rho \equiv -a^\alpha_{\epsilon\rho} \pi^\epsilon \quad \mathrm{mod}\{\theta, \omega\}.$$

Finally, from

$$0 = dF^{\lambda}\Big|_M = \frac{\partial F^{\lambda}}{\partial p^{\alpha}_{\rho}} dp^{\alpha}_{\rho}\Big|_M + (\text{terms in } dz^{\alpha}, dy^{\rho})$$

we obtain symbol relations (11) where

$$r^{\lambda\rho}_{\alpha} = \frac{\partial F^{\lambda}}{\partial p^{\alpha}_{\rho}} .$$

This motivates our terminology.

We note that quasi-linear Pfaffian differential systems look formally like this example; the crucial difference is that in general we do not have the integrability condition

$$d\omega^{\rho} \equiv 0 \mod\{\theta, \omega\}.$$

(15) EXAMPLE: For a manifold Y, the canonical P.D.S. on $G_p(Y)$ is quasi-linear (cf. example (3.6)). It follows that the restriction of this system to any submanifold $M \subset G_p(Y)$ is also quasi-linear. In particular we infer that:

(16) For any Pfaffian differential system (I, J), the prolongations $(I^{(q)}, J^{(q)})$ are quasi-linear for $q \geq 1$.

Comparing with (3.19) we see that, at least theoretically, we may reduce the study of any differential system to that of quasi-linear Pfaffian differential systems.

Before turning to examples arising from isometric embeddings, we want to state the main fact in the theory of quasi-linear systems, namely Cartan's test for involution. To do this we assume that the coframe ω^ρ is chosen generically; we also work at a fixed point $x \in M$ and shall omit reference to it. Finally, we let $\bar{\pi}^\alpha_\rho$ denote the 1-form π^α_ρ considered modulo $\{\theta, \omega\}$, (this has intrinsic meaning, namely $\bar{\pi}^\alpha_\rho$ is the equivalence class of π^α_ρ in $T^*_x M/J_x$), and we define the reduced tableau matrix to be

$$(17) \qquad \bar{\pi} = \left\| \begin{matrix} \bar{\pi}^1_1 & \cdots & \bar{\pi}^1_p \\ \vdots & & \vdots \\ \bar{\pi}^s_1 & \cdots & \bar{\pi}^s_p \end{matrix} \right\| .$$

DEFINITION: The reduced Cartan characters $s'_1, s'_2, \ldots, s'_{p-1}$ are defined inductively by

$$s'_1 + \cdots + s'_k = \left\{ \begin{matrix} \text{number of independent forms in the} \\ \text{1st } k \text{ columns of (17)} \end{matrix} \right\} .$$

We define s'_p by the relation

$$s'_p = n - p - \left[s + s'_1 + \ldots + s'_{p-1}\right].$$

It is easy to see that if the P.D.S. has no Cauchy characteristics, then $s'_1 + \ldots + s'_p$ is the number of independent forms in the reduced tableau matrix. The converse is not true in general as can be seen from examples (31) and (36) below.

From the expositional point of view the Cauchy characteristics are somewhat of a nuisance. As we shall see, from a theoretical viewpoint - i.e., in questions involving involution, prolongation, the characteristic variety, etc. - they may be eliminated. If they are eliminated, then the definition of $s'_1, \ldots, s'_p$ is much more symmetrical and dimension counts come out neater. However, as illustrated by the isometric embedding system, it is frequently geometrically unnatural - even undesirable - to factor out the Cauchy characteristics. Hence, we shall frequently ignore them in theoretical discussions but leave them in when discussing examples.

The origin of these characters is as follows. Suppose that the torsion vanishes identically, and make a change of coframe (7) to annihilate the torsion term $c^{\alpha}_{\rho\sigma}$ in (6). Then it is easy to see that (9) and (11) become

$$(18)\qquad \begin{aligned} &\text{(i)} \quad d\theta^\alpha \equiv -\pi^\alpha_\rho \wedge \omega^\rho \mod\{\theta\} \\ &\text{(ii)} \quad r^{\lambda\rho}_\alpha \pi^\alpha_\rho \equiv 0 \mod\{\theta\}. \end{aligned}$$

A p-plane $E \subset T_xM$ satisfying (i) and (ii) in (1) may, instead of by (4), be given by linear equations

$$(19)\qquad \pi^\alpha_\rho - \ell^\alpha_{\rho\sigma}\omega^\sigma = 0.$$

(Actually, this requires the technical assumption that $\operatorname{span}(\pi^\alpha_\rho) \equiv \operatorname{span}(\pi^\epsilon) \mod\{\theta,\omega\}$, which holds, for example, if there are no Cauchy characteristics. We shall ignore this matter which turns out in the end to be unimportant. Cf. Bryant et al. [1]). The condition

$$d\theta^\alpha\big|_E = 0$$

that (19) be an integral element is clearly equivalent to the symmetry condition

$$(20)\qquad \ell^\alpha_{\rho\sigma} = \ell^\alpha_{\sigma\rho}.$$

The integral elements over $x \in M$ thus form a <u>linear</u> space, whose dimension (assumed as always to be independent of x) we denote by t.

THEOREM: (Cartan's test): We have

(21) $$t \leq s_1' + 2s_2' + \ldots + ps_p',$$

with equality holding if, and only if, (I,J) is involutive.

PROOF: A detailed proof is given in Bryant et al. [1]. As discussed there, from a certain point of view Cartan's test for involution is the central result in the general theory of exterior differential systems. It has a most remarkable reformulation in terms of commutative algebra.

We remark that, by making the special definition of s_p' as above, Cartan's test remains valid even when the system in question has Cauchy characteristics.

We give here a combination of outline of a proof and explanation of why Cartan's test works.

Return for a moment to the case of a differential ideal $\mathcal{I}$ (no independence condition). Let $\varphi^1,\ldots,\varphi^n$ be a local coframe field about x in M. A local coordinate system in the fibre $G_p(M)_x$ is given by (t_ρ^γ), where

(22) $$\varphi^\gamma = t_\rho^\gamma \varphi^\rho,$$

where for the moment we use the index ranges

$$1 \leq \rho, \sigma \leq p, \quad p + 1 \leq \gamma \leq n.$$

The coordinates in $G_p(M)_x$ are valid on the set of p-planes E for which $\varphi^1 \wedge \ldots \wedge \varphi^p \big|_E \neq 0$.

Suppose that E^p is a p-dimensional integral element of $\mathcal{I}$ at x which admits a regular flag (cf. 3.14). We may assume that E^p is given by (22), i.e. that $\varphi^1 \wedge \ldots \wedge \varphi^p \neq 0$. Making a linear change with constant coefficients in $\varphi^1, \ldots, \varphi^p$, if necessary, we may assume that

$$E^k = \{\varphi^{k+1}\big|_{E^p} = 0, \ldots, \varphi^p\big|_{E^p} = 0\}, \quad k = 0, \ldots, p - 1$$

is a regular flag in E^p. Lt $v_1, \ldots, v_p$ be the basis of E^p dual to $\varphi^1\big|_{E^p}, \ldots, \varphi^p\big|_{E^p}$. Thus

$$E^k = \operatorname{span}(v_1, \ldots, v_k), \quad k = 1, \ldots, p - 1.$$

Now E^1 is a 1-dimensional integral element of $\mathcal{I}$ means that v_1 satisfies the s independent linear equations of $\mathcal{I}^{(1)}$. This means that the t_1^γ satisfy s independent linear equations. In fact, $\varphi^\rho(v_1) = \delta_1^\rho$ and $\pi^\gamma(v_1) = t_1^\gamma$, so that if $\theta^\alpha = A_\rho^\alpha \varphi^\rho + B_\gamma^\alpha \varphi^\gamma$,

$\alpha = 1,\ldots,s$ is a basis of $\mathcal{I}^{(1)}$, then

$$0 = \theta^{\alpha}(v_1) = A_1^{\alpha} + \beta_{\gamma}^{\alpha} t_1^{\gamma}$$

are the s independent linear equations the t_1^{γ} must satisfy.

Now $E^2 = \mathrm{span}(v_1, v_2)$ is a 2-dimensional integral element means that v_2 must satisfy the $s + s_1$ independent linear equations of the polar system of E^1. As $\varphi^{\rho}(v_2) = \delta_2^{\rho}$ and $\varphi^{\gamma}(v_2) = t_2^{\gamma}$, these are $s + s_1$ independent linear equations on the t_2^{γ}.

Continuing in this way we see that t_{ρ}^{γ} must satisfy the $s + s_1 + \ldots + s_{\rho-1}$ independent linear equations of the polar system of $E^{\rho-1}$, for $\rho = 1,\ldots,p$. It follows then that in order for E^p to be a p-dimensional integral element which admist a regular flag its coordinates t_{ρ}^{γ} must satisfy $s + (s + s_1) + (s + s_1 + s_2) + \ldots + (s + s_1 + \ldots + s_{p-1}) = ps + (p - 1)s_1 + \ldots + s_{p-1}$ independent linear equations. Thus, as the space of p-dimensional integral elements admitting a regular flag is open in $V_p(\mathcal{I})$,

$$\text{(23)} \qquad p(n - p) - (ps + (p - 1)s_1 + \ldots + s_{p-1}) = \dim \text{ fibre } V_p(\mathcal{I}) \text{ at } x.$$

Let us return now to our quasi-linear P.D.S.

(I,J). For our local coframe field about x we now take $\omega^\rho, \theta^\alpha, \pi^\epsilon$ of (18). Suppose further that E^p is now an integral element of (I,J), that, as above, $v_1, \ldots, v_p$ is the basis of E^p dual to $\omega^1|_{E^p}, \ldots, \omega^p|_{E^p}$, and that $E^k = \mathrm{span}(v_1, \ldots, v_k)$. $\mathcal{I}$ now denotes the differential ideal generated by I.

From (i) of (18) it is easy to see that the polar system of E^1 is

$$\mathrm{span}(\pi_1^\alpha - \pi_\rho^\alpha(v_1)\omega^\rho,\ \theta^\alpha).$$

Its reduced polar system is its polar system modulo $J(x)$ which is

$$\mathrm{span}(\bar{\pi}_1^\alpha).$$

Hence s_1' is the dimension of the reduced polar system of E_1. Clearly

$$s'_1 \leq s_1,$$

because $s + s_1$ = maximum of the dimension of the polar system of a 1-dimensional integral element at x, and thus

$$s_1 \geq \dim \mathrm{span}(\pi_1^\alpha - \pi_\rho^\alpha(v_1)\omega^\rho) = \dim \mathrm{span}(\bar{\pi}_1^\alpha) = s_1'.$$

Note that the first equality here follows from the fact that the ω^ρ are independent on E^p.

The polar system of E^2 is easily computed from (i) of (18) to be

$$\mathrm{span}(\pi_1^\alpha - \pi_\rho^\alpha(v_1)\omega^\rho,\ \pi_2^\alpha - \pi_\rho^\alpha(v_2)\pi^\rho,\ \theta^\alpha),$$

and its *reduced polar system* is this modulo $J(x)$:

$$\mathrm{span}(\bar{\pi}_1^\alpha,\ \bar{\pi}_2^\alpha).$$

Hence

$$s_1' + s_2' = \dim \mathrm{span}(\bar{\pi}_1^\alpha,\ \bar{\pi}_2^\alpha)$$

$$= \dim \mathrm{span}(\pi_1^\alpha - \pi_\rho^\alpha(v_1)\omega^\rho,\ \pi_2^\alpha - \pi_\rho^\alpha(v_2)\omega^\rho) \leq s_1 + s_2.$$

Continuing in this way we find that the *reduced polar system* of E^k is

$$\mathrm{span}(\bar{\pi}_j^\alpha : j = 1,\ldots,k),$$

and consequently

$$s_1' + \ldots + s_k' \leq s_1 + \ldots + s_k, \quad k = 1,2,\ldots,p-1;$$

i.e.

$$s'_k \leq s_k, \quad k = 1, \ldots, p - 1. \tag{24}$$

Combining (23) and (24) with the observation that the space of integral elements of (I, J) is contained in $V_p(\mathcal{I})$, we obtain

$$\begin{aligned} t &\leq (n - p)p - (ps + (p - 1)s_1 + \ldots + s_{p-1}) \\ &\leq (n - p)p - (ps + (p - 1)s'_1 + \ldots + s'_{p-1}), \end{aligned} \tag{25}$$

which gives (21).

If equality holds in (21) then equality must hold throughout (25) and (24), which means that our chosen flag in the integral element E^p is regular. Hence (I, J) is involutive.

Conversely, if (I, J) is involutive, then we pick an integral element E^p of (I, J) which admits a regular flag. It follows that $s'_k = s_k$, $k = 1, \ldots, p - 1$, and that equality holds in (21) by the analysis which led to (23). □

DEFINITION: If $s'_\ell \neq 0$, but $s'_{\ell+1} = \ldots = s'_p = 0$, then we say that (I, J) <u>has character</u> ℓ, and we define the <u>Cartan integer</u> σ by $\sigma = s'_\ell$.

The proof of the Cartan-Kähler Theorem (namely, the application of the Cauchy-Kowaleska Theorem) gives:

(26) The local integral manifolds of an involutive quasi-linear, real analytic P.D.S. of character ℓ depend on σ arbitrary functions of ℓ variables.

This is most concretely manifested in the degree of freedom allowed in the specification of the data for the Cauchy problem of a P.D.S. We illustrate this result with two classical examples.

The <u>wave equation</u>

$$\frac{\partial^2 u}{\partial t^2}(x,t) = \frac{\partial^2 u}{\partial x_1^2}(x,t) + \dots + \frac{\partial^2 u}{\partial x_n^2}(x,t) \ , \ x = (x_1, \dots, x_n).$$

Put $p = \dfrac{\partial u}{\partial t}$, $p_i = \dfrac{\partial u}{\partial x_i}$, $q = \dfrac{\partial^2 u}{\partial t^2}$, $q_i = \dfrac{\partial^2 u}{\partial x_i \partial t}$,

$p_{ij} = \dfrac{\partial^2 u}{\partial x_i \partial x_j} = p_{ji}$, and

$$F(x_i, t, u, p, q, p_i, q_i, p_{ij}) = q - (p_{11} + \dots + p_{nn}).$$

Let $M = \{F = 0\} \subset \mathbb{R}^{(n+1)(n+6)/2}$.

Consider the P.D.S. given locally on M by

$$\text{(i)}\qquad \begin{aligned} \theta &= du - (pdt + p_i dx_i) \\ \theta^{\circ} &= dp - (qdt + q_i dx_i) \\ \theta_i &= dp_i - (q_i dt + p_{ij} dx_j) \end{aligned}$$

$$\text{(ii)}\qquad dx^1 \wedge \dots \wedge dx^n \wedge dt \neq 0.$$

Then

$$\text{(27)}\qquad \begin{aligned} d\theta &\equiv 0 \quad \mathrm{mod}(\theta, \theta^{\circ}, \theta_i) \\ d\theta^{\circ} &= -dq \wedge dt - dq_i \wedge dx_i \\ d\theta_i &= -dq_i \wedge dt - dp_{ij} \wedge dx_j . \end{aligned}$$

The reduced tableau matrix is

$$\begin{vmatrix} 0 & 0 & 0 & \dots & 0 \\ dq & dq_1 & dq_2 & \dots & dq_n \\ dq_1 & dp_{11} & dp_{12} & \dots & dp_{1n} \\ \vdots & \vdots & \vdots & & \vdots \\ dq_n & dp_{n1} & dp_{n2} & \dots & dp_{nn} \end{vmatrix} \quad \mathrm{mod}\{J\}.$$

The symbol relations are:

$$0 = dF = dq - dp_{11} - \dots - dp_{nn} \quad \text{on } M.$$

One easily counts to find

$$s'_k = n + 2 - k\ , \quad k = 1,\dots,n$$
$$s'_{n+1} = 0.$$

Thus

$$(28)\qquad \begin{aligned} &s'_1 + 2s'_2 + \dots + ns'_n + (n+1)s'_{n+1} \\ &= \sum_{k=1}^{n} k(n + 2 - k) = n(n+1)(n+5)/6. \end{aligned}$$

The equations of an admissible $n + 1$ dimensional integral element are $\theta = \theta^{\circ} = \theta_i = 0$ and

$$\begin{aligned} dq &= \ell dt + \ell^i dx_i \\ dq_i &= \ell_i dt + \ell_i^j dx_j \\ dp_{ij} &= \ell_{ij} dt + \ell_{ij}^k dx_k, \quad \ell_{ij} = \ell_{ji},\ \ell_{ij}^k = \ell_{ji}^k. \end{aligned}$$

From the symbol relation and (27) we get

$$\begin{aligned} \ell &= \ell_{11} + \dots + \ell_{nn} \\ \ell^k &= \ell_{11}^k + \dots + \ell_{nn}^k \\ \ell_i &= \ell^i,\ \ell_i^j = \ell_{ij},\ \ell_{ij}^k = \ell_{ik}^j. \end{aligned}$$

Tallying up the number of independent ℓ's we get

$$t = \frac{1}{2}n(n+1) + \begin{bmatrix} n+2 \\ n-1 \end{bmatrix} = n(n+1)(n+5)/6.$$

Comparing this with (28) we see that the P.D.S. is involutive.

Furthermore $\ell = n$, $\sigma = s'_n = 2$, and the general solution depends on 2 arbitrary functions of n variables, which corresponds to the classical result that given functions $h(x)$ and $k(x)$ there is a unique solution $u(x,t)$ of the wave equation satisfying

$$u(x,0) = h(x), \quad \frac{\partial u}{\partial t}(x,0) = k(x).$$

The general non-degenerate 2nd order P.D.E. is worked out in Bryant et al. [1].

The <u>Cauchy-Riemann equations</u>

$$\frac{\partial u}{\partial x^i}(x,y) = \frac{\partial v}{\partial y^i}(x,y), \quad \frac{\partial u}{\partial y^i}(x,y) = -\frac{\partial v}{\partial x^i}(x,y)$$

$$x = (x^1,\ldots,x^n), \quad y = (y^1,\ldots,y^n).$$

Put $p_i = \dfrac{\partial u}{\partial x^i}$, $q_i = \dfrac{\partial v}{\partial x^i}$, $r_i = \dfrac{\partial u}{\partial y^i}$, $s_i = \dfrac{\partial v}{\partial y^i}$, and

$$F^i(x,y,u,v,p,q,r,s) = p_i - s_i$$

$$G^i(x,y,u,v,p,q,r,s) = r_i + q_i, \quad i = 1,\ldots,n.$$

Consider the P.D.S. given locally on

$$M = \{F^i = 0,\ G^i = 0 : i = 1,\dots,n\} \subseteqq \mathbb{R}^{6n+2}$$

by

(i)
$$\begin{aligned} \theta^1 &= du - (p_i dx^i + r_i dy^i) \\ \theta^2 &= dv - (q_i dx^i + s_i dy^i) \end{aligned}$$

(ii)
$$dx^1 \wedge \dots \wedge dx^n \wedge dy^1 \wedge \dots \wedge dy^n \neq 0.$$

Then

(29)
$$\begin{aligned} d\theta^1 &= -dp_i \wedge dx^i - dr_i \wedge dy^i = 0 \\ d\theta^2 &= -dq_i \wedge dx^i - ds_i \wedge dy^i = 0, \end{aligned}$$

and the symbol relations are

$$0 = dF^i = dp_i - ds_i, \quad 0 = dG^i = dr_i + dq_i.$$

The reduced tableau matrix is

$$\begin{bmatrix} dp_i \dots dp_n & dr_1 \dots dr_n \\ dq_1 \dots dq_n & ds_1 \dots ds_n \end{bmatrix} = \begin{bmatrix} dp_1 \dots dp_n & -dq_1 \dots -dq_n \\ dq_1 \dots dq_n & dp_1 \dots dp_n \end{bmatrix}$$

mod$\{\underline{J}\}$.

One easily counts to find

$$s_1' = \ldots = s_n' = 2, \qquad s_{n+1}' = \ldots = s_{2n}' = 0.$$

Thus

$$s_1' + 2s_2' + \ldots + 2ns_{2n}' = n(n+1).$$

The equations of an admissible $2n$ dimensional integral element are $\theta^1 = 0 = \theta^2$ and

$$\begin{aligned} dp_i &= \ell_{ij}dx^j + t_{ij}dy^j \\ dq_i &= m_{ij}dx^j + n_{ij}dy^j, \end{aligned}$$

as $ds_i = dp_i$ and $dr_i = -dq_i$ by the symbol relations. Substituting these into (29) we obtain

$$\begin{aligned} m_{ij} &= -t_{ij} = -t_{ji} \\ n_{ij} &= \ell_{ij} = \ell_{ji}. \end{aligned}$$

Thus the number of independent coefficients is

$$t = n(n+1),$$

and we see that the P.D.S. is involutive.

Furthermore, $\ell = n$, $\sigma = s_n' = 2$, and the general solution depends on 2 arbitrary functions of n variables, which corresponds to the elementary fact that given 2 arbitrary real analytic functions $h(x)$ and $k(x)$ then $h(x) + \sqrt{-1}\, k(x)$ has a holomorphic extension into a domain of $\mathbb{C}^n$ obtained by substituting $z = x + \sqrt{-1}\, y$ into a series expansion of $h(x) + \sqrt{-1}\, k(x)$.

Before considering more examples, remark that if in (17) we have

$$\bar{\pi} = \begin{Vmatrix} 0 & 0 \\ \bar{\varphi} & 0 \end{Vmatrix}, \tag{30}$$

then _in computing the_ s'_ρ _and testing for involution we may ignore the 0-blocks and work only with_ $\bar{\varphi}$. In technical terms, we may ignore the 1st derived system and the Cauchy characteristics of I (cf. Bryant et al. [1]).

(31) EXAMPLE: We reconsider example (3.3). From

$$d\omega^\mu \equiv -\omega^\mu_i \wedge \omega^i \quad \mathrm{mod}\{\omega^\mu\} \tag{32}$$

we infer that the system is quasi-linear with reduced tableau matrix of the form (30) where

$$\bar{\varphi} = \begin{Vmatrix} \omega^{n+1}_1 & \cdots & \omega^{n+1}_n \\ \vdots & & \vdots \\ \omega^{n+r}_1 & \cdots & \omega^{n+r}_n \end{Vmatrix} \quad \mathrm{mod}\{\omega^a, \omega^i_j, \omega^\mu_\upsilon\}.$$

By "eyeballing" this matrix (there are no symbol relations, the ω^μ_i are linearly independent) we see that

$$s'_1 = r, \ldots, s'_n = r$$

(here, we consider only the $\overline{\varphi}$ part of the reduced tableau matrix, and for this part $p = n$). Integral elements are given by

$$\omega^\mu_i - \ell^\mu_{ij}\omega^j = 0 \quad \text{where by (32)}$$

(33)

$$\ell^\mu_{ij} = \ell^\mu_{ji}.$$

It follows that

$$t = rn(n + 1)/2 = s'_1 + 2s'_2 + \ldots + ns'_n.$$

Thus, the system is involutive and the solutions depend on r functions of n variables. This is as expected, since the submanifolds $X^n \subset \mathbf{E}^{n+r}$ may be locally given as graphs by

$$(y^i) \to (y^i; z^\mu(y))$$

where the $z^\mu(y)$ are r arbitrary functions of

$y^1,\ldots,y^n$.

Some comment is in order here to explain why the above counts are correct even though a large number of the variables were omitted from consideration. The reduced tableau matrix is

$$(\bar{\varphi},0)$$

where the 0 is an $r \times (n(n-1)/2 + r(r-1)/2)$ block. An easy count shows that $s'_m = 0$ for $m = n+1,\ldots,p$ (where $p = n + n(n-1)/2 + r(r-1)/2$). Thus

$$s'_1 + 2s'_2 + \ldots + ns'_n = s'_1 + 2s'_2 + \ldots + ps'_p.$$

On the other hand the integral elements are given by

$$\omega^\mu_i = \ell^\mu_{ij}\omega^j + \frac{1}{2}\,\ell^{\mu k}_{ij}\omega^j_k + \frac{1}{2}\,\ell^{\mu\sigma}_{i\upsilon}\omega^\upsilon_\sigma,$$

where $\ell^{\mu k}_{ij} = -\ell^{\mu j}_{ik}$ and $\ell^{\mu\sigma}_{i\upsilon} = -\ell^{\mu\upsilon}_{i\sigma}$. By (32)

$$\ell^\mu_{ij} = \ell^\mu_{ji}\,, \quad \ell^{\mu k}_{ij} = 0 = \ell^{\mu\sigma}_{i\upsilon}.$$

It follows that we still have

$$t = rn(n + 1)/2,$$

and the above abbreviated computation is seen to be correct.

To see why this works, observe that in (i) of (3.3) and in (32) the only forms appearing are

$$\omega^i, \ \omega^\mu, \ \omega^\mu_i. \tag{34}$$

From the structure equations (2.3) for $E(n+r)$ one sees easily that the forms (34), set equal to zero, define a completely integrable system on $E(n+r)$. The maximal integral submanifolds, which have dimension $n(n - 1)/2 + r(r - 1)/2$, are the Cauchy characteristics of (3.3).

On a Cauchy characteristic the forms (34) are all zero. Geometrically, this means that if $(x,e) = (x,e_i,e_\mu)$ is a point in this characteristic, then

$$dx = 0,$$

which means that x is fixed, and

$$de_i = \omega^j_i e_j$$
$$de_\mu = \omega^\nu_\mu e_\nu,$$

which means that this whole characteristic is given by

$$(35) \qquad \{(x,(e_1,\dots,e_n)A,(e_{n+1},\dots,e_{n+r})B): \\ A \in O(n), B \in O(r)\}.$$

In summary, the abbreviated calculation above, made by ignoring the variables ω^i_j and ω^μ_ν, is the result of considering (3.3) on

$$M/\{\text{Cauchy characteristics}\}.$$

In effect, the variables in the Cauchy characteristics are inessential to the problem of finding Darboux subbundles in $E(n+r)$; i.e., of finding integral submanifolds of (3.3).

Thus it is that the P.D.S. reflects the geometric fact that any Darboux frame bundle in $E(n+r)$ (i.e. any integral submanifold of (3.3)) is obtained from any cross-section (x,e) of the Darboux frame bundle over some n-dimensional submanifold $X \subset E^{n+r}$.

As we shall see, the Cauchy characteristics can be ignored not only for questions of involution, but also for the characteristic variety (to be defined in Chapter 6), and for counting the number of effectively different solutions.

We also note from (33) that the integral elements give the 2nd fundamental forms of submanifolds $X^n \subset E^{n+r}$.

The general rule is this: If points of M look like k-jets, then the integral elements of (I,J) look like (k + 1)-jets. This is intuitively clear.

(36) EXAMPLE: We reconsider example (3.24). For the P.D.S. (3.25), which we obtained from the P.D.E. (3.24), we see from (3.26) that its reduced tableau matrix is

$$(dp_1, \ldots, dp_n) \quad \mathrm{mod}\{\underline{J}\}.$$

From (3.27) and the assumption that $F_{p_n} \neq 0$ on M, we have

$$s_1' = \ldots = s_{n-1}' = 1,$$

and

$$s_n' = 2n - n - n = 0.$$

An integral element of (3.25) has equations i) of (3.25) and

$$dp_i = \ell_{ij} dx^j,$$

where by (3.26)

$$\ell_{ij} = \ell_{ji}.$$

Furthermore, by (3.27)

$$F_{x^i} + F_z p_i + F_{p_j} \ell_{ji} = 0.$$

It follows that the space of integral elements at a point of M has dimension

$$t = n(n - 1)/2,$$

which is equal to

$$s'_1 + 2s'_2 + \ldots + ns'_n.$$

Hence (3.25) is involutive and the general solution depends on 1 arbitrary function of $n - 1$ variables, which agrees with the classical fact that for any class C^1 function $h(x^1, \ldots, x^{n-1})$ there is a solution to (3.24)

$$z = u(x^1, \ldots, x^n)$$

satisfying

$$u(x^1,\ldots,x^{n-1},0) = h(x^1,\ldots,x^{n-1}).$$

To be more precise, as $F_{p_n} \neq 0$ on M, we can solve for $p_n = p_n(x^1,\ldots,x^{n-1})$ in the equation

$$F(x^1,\ldots,x^{n-1},0,h,h_{x^1},\ldots,h_{x^{n-1}},p_n) = 0,$$

in a neighborhood of any particular solution of (37). Then

$$(x^1,\ldots,x^{n-1}) \to (x^1,\ldots,x^{n-1},0,h,h_{x^1},\ldots,h_{x^{n-1}},p_n)$$

is an $n-1$ dimensional submanifold N of M, to which the characteristic vector field v of (3.32) is transverse, (because $F_{p_n} \neq 0$). A solution to (3.25) is then given by the union of the integral curves of v passing through N, and from this one obtains the desired solution to (3.24).

The analogy is now clear between the classical method of characteristics and our construction above of the solutions to (3.3) from an arbitrary n-dimensional submanifold of $\mathbf{E}^{n+r}$.

CHAPTER 5

THE ISOMETRIC EMBEDDING SYSTEM

We reconsider the naive isometric embedding system (3.5). Its structure equations are

$$\text{(1)}\quad \begin{aligned} &\text{(i)} \quad d(\omega^i - \bar{\omega}^i) \equiv -(\omega^i_j - \bar{\omega}^i_j)\wedge\omega^j \\ &\text{(ii)} \quad d\omega^\mu \equiv -\omega^\mu_j\wedge\omega^j, \end{aligned}$$

where $\equiv$ denotes congruence modulo the algebraic ideal generated by the Pfaffian forms (i), (ii) in (3.5). As indicated in example (3.18) we do not expect this system to be involutive. We shall now check that this is indeed the case.

For this we note that the reduced tableau matrix is of the form (4.30), and as explained there we shall only deal with the $\bar{\varphi}$-part. Setting

$$\varphi^i_j = \omega^i_j - \bar{\omega}^i_j = -\varphi^j_i,$$

we obtain for the $\bar{\varphi}$-part

$$\begin{vmatrix} 0 & \varphi_2^1 & \cdots & \varphi_n^1 \\ \varphi_1^2 & 0 & \cdots & \varphi_n^2 \\ \vdots & \vdots & & \vdots \\ \varphi_1^n & \varphi_2^n & \cdots & 0 \\ \omega_1^{n+1} & \omega_2^{n+1} & \cdots & \omega_n^{n+1} \\ \vdots & \vdots & & \vdots \\ \omega_1^{n+r} & \omega_2^{n+r} & \cdots & \omega_n^{n+r} \end{vmatrix}$$

where it is understood that we reduce all 1-forms modulo $\{\underline{J}\}$. As the φ_j^i, $i < j$, and ω_j^μ modulo $\{\underline{J}\}$ are linearly independent, we have

$$s_1' = (n - 1) + r, \ s_2' = (n - 2) + r, \ldots, \ s_n' = r$$

$$(3) \Rightarrow s_1' + 2s_2' + \ldots + ns_n'$$
$$= (n - 1)n(n + 1)/6 + rn(n + 1)/2.$$

On the other hand, by (1) and the Cartan lemma, integral elements are given by

$$\text{(i)} \quad \varphi_j^i - a_{jk}^i \omega^k = 0, \quad a_{jk}^i = a_{kj}^i = -a_{ik}^j$$

$$\text{(ii)} \quad \omega_i^\mu - h_{ij}^\mu \omega^j = 0, \quad h_{ij}^\mu = h_{ji}^\mu.$$

As in the proof of (2.17) we must have $a_{jk}^i = 0$, so that

$$t = rn(n + 1)/2.$$

Comparing with (3) we see that, except for the trivial case $n = 1$, the naive isometric embedding system (3.5) fails to be involutive.

Note that there are integral elements over each point, i.e., by (4.8) torsion vanishes identically. The trouble is that the Pfaffian equations (i) in (3.5) imply additional Pfaffian equations not in the original differential ideal.

According to the general theory we must prolong. The space of integral elements of (3.5) is

$$\mathcal{F}(\overline{X}) \times \mathcal{F}(E^{n+r}) \times \mathbb{R}^{rn(n+1)/2}$$

where $\mathbb{R}^{rn(n+1)/2}$ has coordinates $H = (h^{\mu}_{ij})$ (we may think of the H's as potential 2nd fundamental forms). On this space the 1st prolongation of (3.5) is the P.D.S. (I,J) given locally by the Pfaffian equations

$$(4) \qquad \begin{array}{ll} \text{(i)} & \omega^i - \overline{\omega}^i = 0 \\ \text{(ii)} & \omega^{\mu} = 0 \\ \text{(iii)} & \omega^i_j - \overline{\omega}^i_j = 0 \\ \text{(iv)} & \omega^{\mu}_i - h^{\mu}_{ij}\omega^j = 0, \quad h^{\mu}_{ij} = h^{\mu}_{ji}; \end{array}$$

where from now on the independence condition (iii) in (3.5) will be understood. We claim that the structure equations of this system are

$$(5)\quad \begin{aligned} &\text{(i)} && d(\omega^i - \bar{\omega}^i) \equiv 0 \\ &\text{(ii)} && d\omega^\mu \equiv 0 \\ &\text{(iii)} && d(\omega^i_j - \bar{\omega}^i_j) \equiv -\tfrac{1}{2}(R_{ijk\ell} - \gamma(H,H)_{ijk\ell})\omega^k\wedge\omega^\ell \\ &\text{(iv)} && d(\omega^\mu_i - h^\mu_{ij}\omega^j) \equiv -\pi^\mu_{ij}\wedge\omega^j, \end{aligned}$$

all $\bmod\{\underline{I}\}$, where $\pi^\mu_{ij} = \pi^\mu_{ji}$ are given by

$$\pi^\mu_{ij} = dh^\mu_{ij} + \omega^\mu_\nu h^\nu_{ij} - h^\mu_{kj}\omega^k_i - h^\mu_{ki}\omega^k_j.$$

We shall write

$$\pi_{ij} = (\pi^\mu_{ij}) = \pi_{ji},$$

an $\mathbb{R}^r$-valued 1-form.

PROOF: For the first two equations, we simply substitute (iii), (iv) from (4) into (i), (ii) in (1). In fact, this is a special case of a general phenomenon that is worthwhile noting:

(6) The 1st prolongation always contains the original system (these are equations (i), (ii) in (4)). Moreover, the reduced tableau matrix of the 1st prolongation looks like

$$\bar{\pi}^{(1)} = \begin{vmatrix} 0 \\ * \end{vmatrix}$$

where the block of zeroes corresponds to the original system.

Given a sub-bundle $I \subset T^*M$ we define

$$\underline{I}_1 = \{\theta \in \underline{I} : d\theta \equiv 0 \bmod \{\underline{I}\}\}.$$

Assuming that the values $\theta(x) \in I_x$ ($\theta \in \underline{I}_1$ and $x \in M$) give the fibres of a sub-bundle $I_1 \subset I$ (this is just a constant rank assumption), we define the 1st derived system to be the sub-bundle $I_1 \subset I$. We may reformulate (6) by saying that the original system is contained in the 1st derived system of the 1st prolongation; i.e.

$$I \subseteqq (I^{(1)})_1.$$

There is an elaborate theory of derived systems associated to Pfaffian differential systems - cf. Bryant et al. [1].

Continuing now to the proof of (5 iii), we use the Cartan structure equations (2.3) to obtain

$$d(\omega^i_j - \bar{\omega}^i_j) \equiv -\omega^i_k \wedge \omega^k_j - \omega^i_\mu \wedge \omega^\mu_j + \omega^i_k \wedge \omega^k_j - \bar{\Omega}^i_j,$$

by (iii) in (4),

$$\equiv h^\mu_{ik} h^\mu_{j\ell} \omega^k \wedge \omega^\ell - \bar{\Omega}^i_j,$$

by (iv) in (4). Referring to (2.9) and (2.21) we obtain (iii).

Finally, (iv) follows from the structure equations (2.3) together with (4). □

Now we arrive at an interesting situation. Applying the theory in Chapter 4 to equations (iii) and (iv) in (5), we see that:

<u>the torsion of the system</u> (4) <u>is given by</u>

$$\frac{1}{2}\,(\gamma(H,H)_{ijk\ell} - R_{ijk\ell})\omega^k\wedge\omega^\ell\ ,$$

<u>modulo nothing</u>.

Consequently, equating the torsion to zero is equivalent to looking at the subset of $\mathcal{F}(\overline{X}) \times \mathcal{F}(E^{n+r}) \times \mathbb{R}^{rn(n+1)/2}$ where the Gauss equations (cf. 2.21))

$$(7) \qquad \gamma(H,H) = R$$

are satisfied. We let M be an open set, to be specified later, of smooth points of this locus. If the Gauss equations are not smoothly solvable over $\overline{X}$, then $M = \phi$; by a dimension count this will generally be the case if r is small.

On M the $\mathbb{R}^r \otimes \mathbb{R}^{n(n+1)/2}$-valued 1-form $\pi = (\pi^\mu_{ij})$ is subject to the symbol relations that arise by differentiating the Gauss equations (7). To write this neatly, we let

$$\gamma(H,G) = \frac{1}{2}\,(\gamma(H+G,H+G) - \gamma(H,H) - \gamma(G,G))$$

be the symmetric bilinear form obtained by polarizing the quadratic form $\gamma(H,H)$. We then obtain an $S^2(\Lambda^2\mathbb{R}^{n*})$-valued 1-form $\gamma(\pi,H)$, whose entries are

$$2\gamma(\pi,H)_{ijk\ell} = h^\mu_{ik}\pi^\mu_{j\ell} + h^\mu_{j\ell}\pi^\mu_{ik} - h^\mu_{i\ell}\pi^\mu_{jk} - h^\mu_{jk}\pi^\mu_{i\ell}.$$

If we define an $S^2(\Lambda^2\mathbb{R}^{n*})$-valued 1-form DR by

$$(DR)_{ijk\ell} =$$
$$dR_{ijk\ell} - R_{mjk\ell}\bar{\omega}^m_i - R_{imk\ell}\bar{\omega}^m_j - R_{ijm\ell}\bar{\omega}^m_k - R_{ijkm}\bar{\omega}^m_\ell$$

then the exterior differential of (7) gives

$$2\gamma(\pi,H) = DR.$$

(The calculation is elementary: write out (7) explicitly and differentiate, then use (7) and the definitions of π^μ_{ij} and DR).

It is a well-known fact from Riemannian geometry

that

$$(DR)_{ijk\ell} = R_{ijk\ell m}\bar{\omega}^m,$$

where the $R_{ijk\ell m}$ are functions on $\mathcal{F}(\bar{X})$ satisfying the curvature symmetries in the first four indices and the 2nd Bianchi identity in the last three indices.

DEFINITION: The <u>isometric embedding system</u> for $\bar{X}$ in E^{n+r} is defined to be the restriction to M of the Pfaffian differential system (I,J) defined by (4) with the Gauss equations (7) imposed. (Notice that restricting to M imposes the Gauss equations).

In addition to (4), the structure equations of the isometric embedding system, (I,J) restricted to M, are:

$$(8)\qquad \begin{array}{ll} \text{(i)} & d(\omega^i - \bar{\omega}^i) \equiv 0 \\ \text{(ii)} & d\omega^\mu \equiv 0 \\ \text{(iii)} & d(\omega^i_j - \bar{\omega}^i_j) \equiv 0 \\ \text{(iv)} & d(\omega^\mu_i - h^\mu_{ij}\omega^j) \equiv -\pi^\mu_{ij}\wedge\omega^i, \\ \text{(v)} & 2\gamma(\pi,H) = DR \\ \text{(vi)} & \pi^\mu_{ij} = {}^{\mu}_{ji} \end{array} \left.\vphantom{\begin{array}{l} a \\ b \end{array}}\right\} \text{(symbol relations)}$$

For later use, we note that the only non-zero block in

the reduced tableau matrix of the isometric embedding system corresponds to (iv) in (8). All of the important structure of the system will come off (iv), (v), and (vi) in (8).

We are now ready to prove the

(9) THEOREM: The isometric embedding system is involutive if, and only if

$$r \geq n(n-1)/2.$$

In case $r = n(n-1)/2$ its reduced Cartan characters are

$$s'_p = n(n-p)(n-p+1)/2, \quad p = 1,\ldots,n,$$

and thus (I,J) has character $n-1$ and Cartan integer $s'_{n-1} = n$.

(10) COROLLARY (Cartan-Janet). In the real analytic case, local isometric embeddings

$$\overline{X}^n \to E^{n(n+1)/2}$$

exist. They depend on n functions of $n-1$ variables.

We remark that our approach to Cartan-Janet is technically somewhat different from the usual ones. For example, it also gives the intuitively clear result that the isometric embedding system fails to be involutive if $r < n(n - 1)/2$. More importantly, we shall see in the next section that this formulation leads directly to the characteristic variety.

Finally we note that it may seem natural to only "partially prolong" (3.5) by adding to it the Pfaffian equations

$$\omega^i_j - \bar{\omega}^i_j = 0$$

which gives the P.D.S. consisting of (i)-(iii) in (4). In fact, it was this system which Cartan considered in [1] and [2]. A major difference between this system and ours is that this one is not quasi-linear. Another difference is that in the application of Cartan's test to our system the covariant derivatives of the curvature enter in, while only the curvature enters into the test of this partially prolonged system.

PROOF OF THEOREM (9): We shall assume that

$$r \geq n(n - 1)/2;$$

the proof that the system fails to be involutive when $r < n(n-1)/2$ may be done by a similar but simpler computation.

We shall introduce the following vector spaces

$$W = \mathbb{R}^r$$
$$V = \mathbb{R}^n$$
$$K \subset S^2(\Lambda^2 V^*)$$
$$K^{(1)} \subset K \otimes V^*$$

where K is the space of curvature-like tensors $R = \{R_{ijk\ell}\}$ satisfying the familiar symmetries

$$R_{ijk\ell} = -R_{jik\ell} = -R_{ij\ell k}$$
$$R_{ijk\ell} = R_{k\ell ij}$$
$$R_{ijk\ell} + R_{i\ell jk} + R_{ik\ell j} = 0 \quad (1^{st} \text{ Bianchi})$$

and $K^{(1)}$ are those $\{R_{ijk\ell m}\}$ satisfying the curvature symmetries in i, j, k, ℓ plus the 2nd Bianchi identity

$$R_{ijk\ell m} + R_{ijmk\ell} + R_{ij\ell mk} = 0.$$

The left side of the Gauss equations (7) may be viewed as a quadratic mapping

$$\gamma : W \otimes S^2 V^* \to K$$
$$(h^\mu_{ij}) \to (h^\mu_{ik} h^\mu_{j\ell} - h^\mu_{i\ell} h^\mu_{jk})$$

whose differential at $H \in W \otimes S^2V^*$ will be denoted by

$$\gamma_H : W \otimes S^2V^* \to K$$
$$\gamma_H = h^\mu_{ik} dh^\mu_{j\ell} + h^\mu_{j\ell} dh^\mu_{ik} - h^\mu_{i\ell} dh^\mu_{jk} - h^\mu_{jk} dh^\mu_{i\ell};$$

this is a linear mapping. Observe that

$$\gamma_H(G) = 2\gamma(H,G).$$

We shall use the following dimension counts, which may be done directly or found in Kaneda-Tanaka [1] or in §V of Berger-Bryant-Griffiths [2]:

$$\text{(11)}\quad \begin{aligned} &\text{(i)} && \dim K = n^2(n^2 - 1)/12 \\ &\text{(ii)} && \dim K^{(1)} = n^2(n^2 - 1)(n + 2)/24 \\ &\text{(iii)} && \dim W \otimes S^3V^* = \frac{rn(n + 1)(n + 2)}{6} . \end{aligned}$$

Remark that $K \subset \otimes^4V^*$ and $K^{(1)} \subset \otimes^5V^*$ are each irreducible $GL(V)$-modules, but in the case when V has a Euclidean structure they are not irreducible $O(V)$-modules (loc. cit.).

DEFINITION: $H \in W \otimes S^2V^*$ is ordinary if there exists a basis $v_1, \ldots, v_n$ for V such that the vectors

$$H_{\rho\sigma} = H_{\sigma\rho} = H(v_\rho, v_\sigma) \in W, \quad 1 \le \rho,\sigma \le n - 1,$$

are linearly independent. We denote by $\mathcal{H} \subset W \otimes S^2V^*$ the (non-empty since $r \geq n(n-1)/2$) Zariski open set of ordinary H's.

REMARK: To see better the set $\mathcal{H}$ let e_i be a basis of V, w_μ a basis of W. Then $H \in W \otimes S^2V^*$ is given by

$$H(e_i, e_j) = h^\mu_{ij} w_\mu, \quad h^\mu_{ij} = h^\mu_{ji}.$$

Any other basis of V is given by

$$v_i = T^k_i e_k, \quad T = (T^k_i) \in GL(n;\mathbb{R}).$$

Thus

$$H(v_\rho, v_\sigma) = T^i_\rho T^j_\sigma h^\mu_{ij} w_\mu, \quad 1 \leq \rho, \sigma \leq n-1.$$

It follows that $H \notin \mathcal{H}$ if, and only if,

$$\bigwedge_{\rho \leq \sigma} T^i_\rho T^j_\sigma h^\mu_{ij} w_\mu = 0, \quad 1 \leq \rho, \sigma \leq n-1,$$

for all $T \in GL(n;\mathbb{R})$. As $r \geq n(n-1)/2$, this is a finite collection of polynomial equations of $GL(n;\mathbb{R})$ with coefficients polynomials in h^μ_{ij}. To be identically zero on $GL(n;\mathbb{R})$ requires that these coefficients be identically zero. These are the equations of the complement of $\mathcal{H}$.

The main computational step in the proof of Theorem (9) is given by the

(12) PROPOSITION: $\gamma:\mathcal{H} \to K$ is surjective and is everywhere of maximal rank. In particular, the Gauss equations are locally smoothly solvable by ordinary candidates H for a 2nd fundamental form.

PROOF OF THE PROPOSITION: We shall use the linearized Gauss equations in the form:

$$G = \{G_{ij}\} \in \ker \gamma_H \text{ if, and only if,}$$

(13) $$H_{ik}\cdot G_{j\ell} + H_{j\ell}\cdot G_{ik} - H_{i\ell}\cdot G_{jk} - H_{jk}\cdot G_{i\ell} = 0.$$

Here G_{ij}, $H_{k\ell} \in W$.

To prove that γ_H has maximal rank at $H \in \mathcal{H}$ it will suffice by (i) in (11) to show that

(14) $$\dim \ker \gamma_H \leq rn(n + 1)/2 - n^2(n^2 - 1)/12.$$

In fact, under our assumption $r \geq n(n - 1)/2$, γ_H being of maximal rank is equivalent to γ_H being surjective. Thus we must estimate the number of linearly independent solutions to (13).

For this we define mappings, for $1 \leq p \leq n$,

$$\mu_1 : \ker \gamma_H \to W \oplus \ldots \oplus W \quad (n \;\text{ factors})$$

$$\vdots$$

$$\mu_p : \ker \gamma_H \to W \oplus \ldots \oplus W \quad (f(p) \;\text{ factors})$$

$$\vdots$$

where $f(p) = n + (n - 1) + \ldots + (n - p + 1)$ and

$$\mu_1(G) = (G_{11}, G_{12}, \ldots, G_{1n})$$

$$\vdots$$

$$\mu_p(G) = \mu_{p-1}(G) \oplus (G_{p,p}, G_{p,p+1}, \ldots, G_{p,n})$$

$$\vdots$$

Intuitively, if we view G as a symmetric matrix with entries in W, then μ_p maps $\ker \gamma_H$ to the entries in the shaded part of the matrix

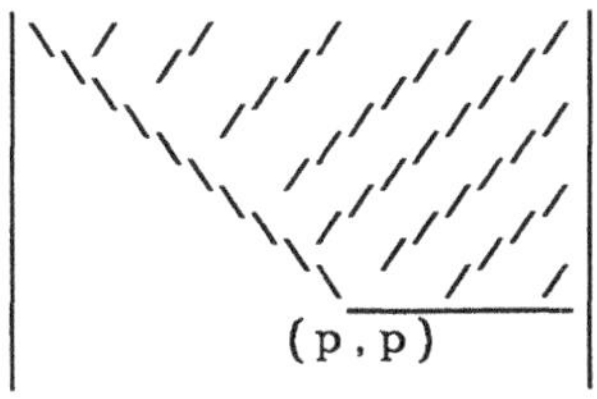

As G is symmetric it is clear that

$$\text{(15)} \qquad \dim \ker \gamma_H = \dim \text{image } \mu_n .$$

Define $t_1, \ldots, t_n$ by

$$t_1 + \ldots + t_p = \dim \text{image } \mu_p .$$

By (14), (15) we must show that

$$t_1 + \ldots + t_n \leq rn(n+1)/2 - n^2(n^2 - 1)/12.$$

Using an elementary calculuation this will follow from the estimate

$$\text{(16)} \qquad t_p \leq (n - p + 1)(r - n(p - 1)/2),$$

which is what we shall now prove.

Among the relations (13) are the following that express the $G_{p\ell}$, $p \leq \ell \leq n$, in terms of the G_{ij} where $1 \leq i \leq p - 1$:

$$\text{(17)} \qquad H_{ik} \cdot G_{p\ell} = H_{i\ell} \cdot G_{pk} + H_{pk} \cdot G_{i\ell} - H_{p\ell} \cdot G_{ik}$$

where $1 \leq i \leq k \leq p - 1$, $\ell \geq p$ and

(18) $H_{ik} \cdot G_{p\ell} - H_{i\ell} \cdot G_{pk} = -H_{p\ell} \cdot G_{ik} + H_{pk} \cdot G_{i\ell}$

where $i < p \le k < \ell$. As H is ordinary there are

$$(p(p - 1)/2)(n - p + 1)$$

independent relations (17) and

$$(p - 1)(n - p + 1)(n - p)/2$$

additional independent relations (18). These two numbers add up to

$$n(n - p + 1)(p - 1)/2.$$

For fixed $\mu_{p-1}(G)$, it follows that there are <u>at most</u>

$$r(n - p + 1) - n(n - p + 1)(p - 1)/2 =$$
$$(n - p + 1)(r - n(p - 1)/2)$$

choices for the $G_{p\ell}$, $p \le \ell \le n$. This is (16). (Note: the "at most" is because there may be additional equations beyond (17), and (18) - in fact, this will turn out not to be the case.)

It remains to show that

(19) $$\gamma(\mathcal{H}) = K.$$

If we can show that there is an $H \in \mathcal{H}$ with

(20) $$\gamma(H,H) = (0),$$

then (19) will follow because γ_H surjective implies that $\gamma(\mathcal{H})$ will then contain a neighborhood of 0 in K, and thus $\gamma(\mathcal{H}) = K$ as $\gamma(\lambda H, \lambda H) = \lambda^2 \gamma(H,H)$.

We construct an $H \in \mathcal{H}$ which satisfies (20). As $r = \dim W \geq n(n-1)/2$, we can choose vectors $H_{\rho\sigma} = H_{\sigma\rho} \in W$, $1 \leq \rho, \sigma \leq n-1$, such that

(21)
- (i) $H_{\rho\rho} \cdot H_{\sigma\sigma} = H_{\rho\sigma} \cdot H_{\rho\sigma} = 1$ if $\rho \neq \sigma$,
- (ii) $\mathrm{span}(H_{\rho\rho}) \perp \mathrm{span}(h_{\sigma\tau} : 1 \leq \sigma < \tau < n)$,
- (iii) $\{H_{\rho\sigma} : 1 \leq \rho < \sigma \leq n-1\}$ is an orthonormal set.

Setting $H_{in} = H_{ni} = 0$ for $1 \leq i \leq n$, we obtain an element $H \in \mathcal{H}$ which satisfies (20).

This completes the proof of (12). We can now complete the proof of (9).

We define

$$M \subset \mathcal{F}(\overline{X}) \times \mathcal{F}(E^{n+r}) \times \mathcal{H}$$

to be the ordinary solutions of the Gauss equations (7). That is,

$$M = \{(\bar{x}, \bar{e}_i; x, e_i, e_\mu; H) : \gamma(H,H) - R(\bar{x}) = 0\}.$$

By Proposition 12, M is a smooth manifold submersing onto $\mathcal{F}(\bar{X}) \times \mathcal{F}(\mathbf{E}^{n+r})$; in particular,

$$\bigwedge_i \omega^i \bigwedge_{i<j} \omega^i_j \bigwedge_{\mu<\nu} \omega^\mu_\nu \neq 0$$

on M. To complete the proof of Theorem (9) we will show that:

(22) Cartan's test is satisfied for the isometric embedding system (I,J) given by (4) restricted to M.

The essential part of the reduced tableau matrix is (at a given point)

$$\bar{\pi} = \begin{vmatrix} \bar{\pi}_{11} & \cdots & \bar{\pi}_{1n} \\ \vdots & & \vdots \\ \bar{\pi}_{n1} & \cdots & \bar{\pi}_{nn} \end{vmatrix}$$

where

$$\bar{\pi}_{ij} = \pi_{ij} \mod\{\underline{J}\},$$

and the π_{ij} are the W-valued 1-forms defined in (iv) of (5).

Here we are considering only what may be called the "$\overline{\varphi}$-part" of the reduced tableau matrix (cf. (4.23)). As previously remarked, Cartan's test is verified for the $\overline{\varphi}$-part if, and only if, it is verified for the whole part of the reduced tableau matrix. This is what is meant by the above phrase "essental part of the reduced tableau matrix".

By (8) the symbol relations are

(23)
(i) $\overline{\pi}_{ij} = \overline{\pi}_{ji}$

(ii) $\gamma(\overline{\pi}, H) = 0$, (as $DR \equiv 0 \mod\{\underline{J}\}$).

Using the proof of proposition (12), combined with the observations that $\overline{\pi} = {}^t\overline{\pi}$ by (i) of (23) and that the relations (ii) in (23) are the same as (13) with $\overline{\pi}$ replacing G, we conclude that

$$s_1' + \ldots + s_p' = t_1 + \ldots + t_p .$$

Thus, by (16),

(24) $s_p' = t_p \leq (n - p + 1)(r - n(p - 1)/2)$.

It follows that

$$(25) \qquad s_1' + 2s_2' + \ldots + ns_n' \leq \frac{rn(n+1)(n+2)}{6} - \frac{n^2(n^2-1)(n+2)}{24}.$$

On the other hand, the vector space $\mathcal{V}$ of integral elements lying over a fixed point $(\bar{x}, \bar{e}_i, x, e_i, e_\mu, H)$ of M is given by linear equations

$$\pi_{ij} - F_{ijk}\omega^k = 0$$

where by (iv) in (8) $F \in W \otimes S^3V^*$, and by (v) in (8)

$$(26) \qquad H_{ik} \cdot F_{j\ell m} + H_{j\ell} \cdot F_{ikm} - H_{i\ell} \cdot F_{jkm} - H_{jk} \cdot F_{i\ell m} = R_{ijk\ell m} \in K^{(1)}.$$

Consider the linear map

$$L_H : W \otimes S^3V^* \to K^{(1)},$$

where $L_H(F)$ is given by the left side of (26). We denote $\{R_{ijk\ell m}\} \in K^{(1)}$ by ∇R. Then

$$\mathcal{V} = \{F \in W \otimes S^3V^* : L_H(F) = \nabla R\}.$$

If $\mathcal{V} \neq \phi$, then

(27) $$t = \dim \mathscr{V} = \dim \ker L_H.$$

We will show that $\mathscr{V} \neq \phi$ by showing that L_H is surjective, and this is proved by showing that

(28) $$\dim \ker L_H \leq \dim W \otimes S^3V^* - \dim K^{(1)}.$$

(Note: the right side of (28) is positive when $r \geq n(n - 1)/2$.)

Assuming (28) for the moment, we conclude first of all that (28) must be an equality. Thus from (25), (27), (28) and (ii) and (iii) of (11) we have

(29) $$s_1' + 2s_2' + \ldots + ns_n' \leq t$$

Hence by Cartan's test (29) must be an equality, and we have proved that the isometric embedding system is involutive when $r \geq n(n - 1)/2$. Furthermore, there must be equality in (25) and (16), which verifies the remark that there are not relations beyond (17), (18) among the G_{ij}, and verifies the formulas for s_p' in the statement of Theorem (9).

To prove (28), we consider $F \in \ker L_H$. For each fixed p put $F_p = \{F_{ijp}\} \in \ker \gamma_H$. The dimension of $\ker L_H$ will be the dimension of $A_1 \oplus \ldots \oplus A_n$, where

$$A_p = \{F_{ijp} : p \leq i \leq j \leq n\} \subseteqq W \oplus \ldots \oplus W$$
$$((n - p + 1)(n - p + 2)/2 \quad \text{factors})$$

for $p = 1,\ldots,n$. (Note: for each p, if $i < p$ or $j < p$, then F_{ijp} occurs as an entry of A_i or A_j). From the proof of Proposition (12)

$$\dim A_p = t_p + \ldots + t_n, \quad p = 1,\ldots,n.$$

Hence

$$\dim \ker L_H = \sum_1^n p t_p,$$

which, with (16) and (ii), (iii) of (11), gives (28). □

CHAPTER 6

THE CHARACTERISTIC VARIETY

We consider a quasi-linear P.D.S. (I,J) on a manifold M, and we set

(1)
$$\begin{aligned} V^* &= J/I \\ \mathbf{P}V^* &= \text{projective bundle associated to } V^* \\ V^*_{\mathbb{C}} &= \text{compleixification of } V^* \\ \mathbf{P}V^*_{\mathbb{C}} &= \text{projective bundle associated to } V^*_{\mathbb{C}}. \end{aligned}$$

Thus, $V^*_x = J_x/I_x$ and $V^* = \bigcup_{x \in X} V^*_x$, $\mathbf{P}V^* = \bigcup_{x \in X} \mathbf{P}V^*_x$, etc; and

$$V^*_x = \operatorname{span}(\bar{\omega}^1, \ldots, \bar{\omega}^p) \quad \text{at } x,$$

where $\omega^1 \wedge \ldots \wedge \omega^p \neq 0$ is the independence condition and $\bar{\omega}^\rho = \omega^\rho \mod\{\underline{I}\}$. We will omit the bars.

We shall define the characteristic variety

$$\Xi \subset \mathbf{P}V$$

of (I,J) in coordinates, and then note that it is

independent of choices.

Characteristic varieties of linear P.D.E. systems live in projectivized cotangent spaces. This is why we have set $J/I = V^*$ instead of setting it equal to V. In other words, $V^* \to M$ is a vector bundle whose restriction to any integral manifold $N \subset M$ is naturally identified with T^*N; the characteristic variety $\Xi \subset \mathbf{P}V^*$ then induces a characteristic variety $\Xi_N \subset \mathbf{P}T^*N$ for each integral manifold N. Here, it may be useful to recall that the defintion of the characteristic variety of a non-linear P.D.E. system depends on being at a particular solution to the system.

Give (I,J) in the form

$$(2)\qquad \begin{array}{llll} \text{(i)} & \theta^\alpha = 0 & & \alpha = 1,\dots,s \\ \text{(ii)} & d\theta^\alpha \equiv -\pi^\alpha_\rho \wedge \omega^\rho & \text{mod}\{\theta\} & \rho = 1,\dots,p \\ \text{(iii)} & r^{\lambda\rho}_\alpha \pi^\alpha_\rho \equiv 0 & \text{mod}\ \{\theta,\omega\} & \lambda = 1,\dots,Q \end{array}$$

where (iii) are the symbol relations. For each

$$\xi = \xi_\rho \omega^\rho \in J/I = V^*, \quad \xi \neq 0,$$

we let

$$[\xi] = [\xi_1,\dots,\xi_p] \in \mathbf{P}V^*$$

be the corresponding point, and define the $Q \times s$ symbol matrix by

$$(3) \qquad \sigma_\xi(x) = \left\| r_\alpha^{\lambda\rho}(x)\xi_\rho \right\| .$$

The characteristic variety

$$\Xi = \bigcup_{x \in X} \Xi_x$$

will be the union of the characteristic varieties

$$(4) \qquad \Xi_x \subset \mathbf{P}V_x^*$$

lying over each point $x \in X$, and for these we have the

DEFINITION: The variety (4) is defined by

$$\Xi_x = \{[\xi] \in \mathbf{P}V_x^* : \sigma_\xi(x) \text{ fails to be injective}\} .$$

Referring to example (4.14), we see that our definition extends the classical notion of the characteristic variety of a P.D.E. system.

To define the complex characteristic variety

$$\Xi_{\mathbb{C}} \subset \mathbf{P}V_{\mathbb{C}}^*$$

we simply repeat the defintion with $V^*_{\mathbb{C}}$ replacing V^*. Clearly Ξ_x is defined by homogeneous polynomial equations. For example, when $s \leq Q$ these are

$$\text{all } s \times s \text{ minors of } \sigma_\xi(x) = 0,$$

and $\Xi_{\mathbb{C},x}$ consists of the complex solutions to these equations. (If $s > Q \geq 1$, then $\Xi_x = \mathbf{P}V^*_x$. If $Q = 0$, then $\Xi_x = \phi$).

Actually, what is really needed to do the theory right is the characteristic sheaf. Without becoming formal, let us think of a sheaf as a vector bundle with possibly jumping fibre dimension. Then the characteristic sheaf

$$\mathcal{M} = \bigcup_{(x,[\xi])\in \mathbf{P}V^*_{\mathbb{C}}} \mathcal{M}_{x,[\xi]}$$

where

$$\mathcal{M}_{x,[\xi]} = (\ker \sigma_\xi(x))^* \subset I_x.$$

It is a sheaf over $\mathbf{P}V^*_{\mathbb{C}}$ whose support is $\Xi_{\mathbb{C}}$; what it does is keep track not only of where the symbol fails to be injective but also by how much.

As motivation for this notion we mention the following result (cf. Bryant et al. [1]):

(5) The characteristic sheaf (but not the

characteristic variety) of an involutive P.D.S. uniquely determines the symbol mappings (3), up to adding trivial symbols.

Characteristic varieties of differential systems have a number of remarkable properties, and as shown by examples they constitute a very rich invariant. Certainly, in practice when given a P.D.S. arising from a geometric problem the first thing one now does is to determine its characteristic variety. For the purposes of these talks we shall only need four properties of Ξ, three of which we now explain; amplifications and proofs may be found in loc. cit.

For the first, we shall say that a P.D.S. (I,J) is of <u>finite type</u> if a prolongation $(I^{(q)},J^{(q)})$ is either incompatible (i.e., there are no integral elements), or is a Frobenius system (cf. (3.15)). Systems of finite type have the property that their solutions depend at most on constants; geometrically we may express this as follows:

(6) If (I,J) is of finite type, then there is an integer q such that if two connected integral manifolds N, N' osculate to order q at one point $x \in M$, then $N = N'$.

This q and the one above are the same.

The basic result concerning finite type systems is:

(7) If the complex characteristic variety

$$\Xi_{\mathbb{C}} = \phi,$$

then (I,J) is of finite type.

As the ordinary Laplace equation shows, this result is false if we try to use the real characteristic variety and not the complex one.

Recently, systems of finite type have been frequently turning up (cf. the examples and references in Bryant et al. [1]).

(8) EXAMPLE: We show that the complex characteristic variety of a Frobenius system (I,J) is empty. A Frobenius system is given locally by (2) with

$$\pi^{\alpha}_{\rho} = 0 \qquad \forall\ \alpha,\rho.$$

Thus the symbol relations are indexed by pairs (α,ρ):

$$r^{(\alpha\rho)\sigma}_{\beta}\pi^{\beta}_{\sigma} = \pi^{\alpha}_{\rho} = 0 \qquad \beta = 1,\ldots,s;\ \rho = 1,\ldots,p;$$

which give

$$r_{\beta}^{(\alpha\rho)\sigma} = \delta_{\beta}^{\alpha}\delta^{\rho\sigma}.$$

For any $[\xi] \in PV_{\mathbb{C}}^{*}$, the symbol matrix σ_{ξ} is singular if there exists a non-zero $w = (w^{\beta}) \in \mathbb{C}^{s}$ such that

$$0 = r_{\beta}^{(\alpha\rho)\sigma}\xi_{\sigma}w^{\beta} = \xi_{\rho}w^{\alpha}, \quad \forall\ \alpha,\rho.$$

As $\xi \neq 0$, we must have $w = 0$; i.e.,

$$\sigma_{\xi}:\mathbb{C}^{s} \to \mathbb{C}^{sp}$$

is injective for every $[\xi] \in PV_{\mathbb{C}}^{*}$, i.e., $\Xi_{\mathbb{C}} = \phi$.

(9) EXAMPLE: We compute the symbol matrix for a P.D.S. which is partly like a Frobenius system. The result is that the symbol matrix decomposes into a direct sum, one part of which is never non-singular and hence which may be ignored in determining the characteristic variety. This example is directly relevant to our computation later on of the characteristic variety of the isometric embedding system.

We suppose (I,J) is given locally as

$$
(10)\qquad
\begin{array}{lll}
\text{(i)} & \theta^{a} = 0 & a,b = 1,\dots,s_1 \\
\text{(ii)} & \theta^{\alpha} = 0 & \alpha,\beta = s_1+1,\dots,s_1+s_2 \\
\text{(iii)} & d\theta^{a} \equiv 0 \mod\{\underline{I}\} & \\
\text{(iv)} & d\theta^{\alpha} \equiv -\pi^{\alpha}_{\rho}\wedge\omega^{\rho} \mod\{\underline{I}\} & \sigma,\rho = 1,\dots,p \\
\text{(v)} & r^{\lambda\rho}_{\alpha}\pi^{\alpha}_{\rho} \equiv 0 \mod\{\underline{J}\} & \lambda = 1,\dots,Q.
\end{array}
$$

Now (v) gives some of the symbol relations, but we get in addition, from (iii), the symbol relations analogous to (8):

$$0 = \pi^{a}_{\rho} = r^{(a\rho)\sigma}_{b}\pi^{b}_{\sigma}; \quad \text{i.e.}$$

$$r^{(a\rho)\sigma}_{b} = \delta^{a}_{b}\delta^{\rho\sigma}.$$

For any $[\xi] \in PV^{*}_{\mathbb{C}}$, the symbol matrix is

$$
(11)\qquad \sigma_{\xi} = \begin{vmatrix} r^{(a\rho)\sigma}_{b}\xi_{\sigma} & 0 \\ 0 & r^{\lambda\rho}_{\alpha}\xi_{\rho} \end{vmatrix}.
$$

As we saw in example (8), the upper left $s_1p \times s_1$ block of σ_{ξ} is non-singular for every $\xi \neq 0$. Consequently, σ_{ξ} is singular if, and only if, the lower right $r \times s_2$ block

$$
(12)\qquad (r^{\lambda\rho}_{\alpha}\xi_{\rho})
$$

is singular. We will call (12) the essential part of the symbol of a system of the type of (10). For brevity we shall simply call the essential part σ_ξ.

(13) EXAMPLE: To explain a little why (7) should be true, consider a homogeneous constant coefficient P.D.E. system for one unknown function

(14) $$\sum_{|I|=m} a^{\lambda I} \frac{\partial^m u}{\partial x^I} = 0 , \qquad \lambda = 1,\ldots,Q.$$

The symbol "matrix" in the P.D.E. theory sense is the column vector

$$\sigma(\xi) = {}^t(P^1(\xi),\ldots,P^Q(\xi))$$

where

$$P^\lambda(\xi) = \sum_I a^{\lambda I}\xi_I, \qquad (\xi_I = \xi_{i_1}\cdots\xi_{i_m}).$$

The complex characteristic variety is given by

$$\{[\xi] \in \mathbb{C}P^{n-1} : P^1(\xi) = \ldots = P^Q(\xi) = 0\}.$$

Suppose this variety is empty. Then by Hilbert's nullstellensatz (this is where we use $\mathbb{C}$) there is an integer q such that every homogeneous form of degree greater than or equal to q is in the ideal generated by the $P^\lambda(\xi)$'s. In particular, we have

$$\xi_J = Q_{J\lambda}(\xi)P^{\lambda}(\xi), \quad |J| \geq q.$$

From (8) this gives

$$\frac{\partial^q u(x)}{\partial x^J} = 0, \quad \text{for all } J \text{ with } |J| \geq q.$$

Consequently, u is a polynomial and the solution space of (14) is a finite-dimensional vector space.

For the second property of the characteristic variety of a P.D.S. we recall from §4 the notion of the character ℓ and Cartan integer σ of a P.D.S. (See just above (4.26)).

(15) Suppose that (I,J) is an involutive P.D.S. of character ℓ and with Cartan integer σ. Then

$$\dim \Xi_{\mathbb{C},x} = \ell - 1$$
$$\deg \Xi_{\mathbb{C},x} = \sigma.$$

Here, $\deg \Xi_{\mathbb{C},x}$ is the degree of the complex projective variety $\Xi_{\mathbb{C},x}$ in the scheme-theoretic sense; i.e., we count multiplicities. Counting multiplicities properly is slightly complicated in that it involves the fibre dimension of the characteristic sheaf over a general point of $\Xi_{\mathbb{C},x}$, but we will not dwell on this matter here.

In practice this result allows us to readily compute ℓ and σ in examples. However, it may also be turned around: Suppose we know that

$$(16)\quad \begin{array}{lll} \text{(i)} & \Xi_{\mathbb{C}.x} \subset \mathbf{P}V^*_{\mathbb{C},x} & \text{is a hypersurface} \\ \text{(ii)} & \deg \Xi_{\mathbb{C},x} & \text{is odd.} \end{array}$$

By (i) we have that $\Xi_x \subset \mathbf{P}V^*_x$ is defined by one homogeneous equation, which by (ii) must have a non-empty set of real zeroes.

If we recall from (5.9) that, for the isometric embedding system in the embedding dimension the character $\ell = n - 1$ and the Cartan integer $\sigma = n$, then we may draw the following conclusion:

(17) For the isometric embedding system (5.4) in the embedding dimension $N(n) = n(n + 1)/2$, the real characteristic variety

$$(18)\qquad \Xi \neq \phi$$

for n odd.

In fact, we will see that the conclusion (18) is true when $n \geq 3$. This means for example that, no matter what positivity the curvature of X may have, the isometric embedding system in the embedding dimension is never elliptic if $n \geq 3$ (cf. remark (68) below).

Finally, there is a general notion of <u>hyperbolicity</u> for a P.D.S. (Yang [1]). Although the general concept involves the characteristic sheaf $\mathcal{M}$, the special cases we shall use may be explained in terms of the characteristic variety alone. To motivate these we consider a single linear P.D.E. for one unknown function

(19) $$P(x,D) = \sum_{|I|\leq m} a^I(x) \frac{\partial^{|I|}u(x)}{\partial x^I} = 0, \qquad x = (x^1,\dots,x^p).$$

where $x = (x^1,\dots,x^p)$ Its principal symbol from P.D.E. theory is

$$P_m(x,\xi) = \sum_{|I|=m} a^I(x)\xi_I .$$

For each x the characteristic variety

$$\Xi_x \subset \mathbf{P}^{p-1}$$

is defined by the equation $P_m(x,\xi) = 0$. The classical

condition that (19) be hyperbolic is this:

(20) For each x there exists $\eta \in \mathbf{P}^{p-1}$ such that for any $\xi \neq \eta$ the polynomial

$$P_m(x, t\eta + \xi)$$

has $m = \deg P_m(\xi)$ real roots.

The roots need not be distinct, but they should at least be single-valued functions of ξ (no branching).

Geometrically, (20) is equivalent to the following condition:

(21) There exists $\eta \in \mathbf{P}^{p-1}$ such that each line through η meets Ξ_x in m real points, none of which is η.

For example, when $p = 3$ and $m = 3$ the principal symbol is a cubic polynomial. It is well-known that, by a real linear change of coordinates in $\mathbf{P}^2$, every smooth cubic may be put into the Weierstrass normal form

(22) $$z^2 = f(y), \quad f(y) = y^3 + ay + b.$$

where $[1,y,z]$ are normalized homogeneous coordinates

in $\mathbf{P}^2$. The cubic defined by (22) looks like either

Figure 1

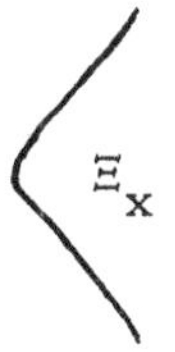

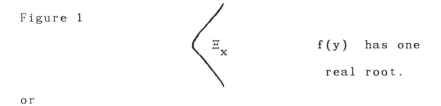

f(y) has one real root.

or

Figure 2

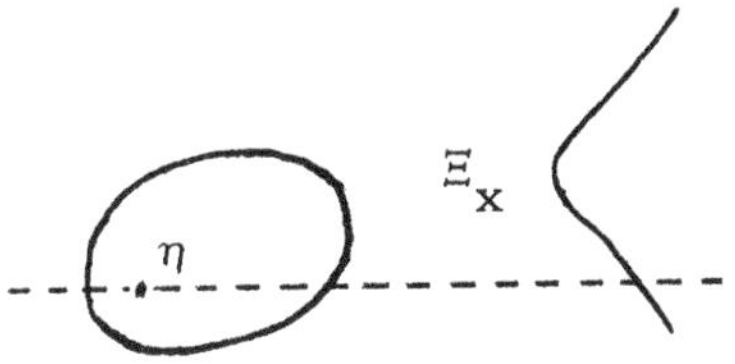

f(y) has three real roots.

The cubic in Fig. 2 is hyperbolic (the vertical line meets Ξ_x in the point at infinity on the z-axis), whereas that in Fig. 1 is not hyperoblic.

In order to define hyperbolicity for a P.D.S. we begin by saying that a hypersurface $\Xi_x \subset \mathbf{P}^{p-1}$ is hyperbolic iff the condition (21) is satisfied.

More generally, suppose that the complex characteristic variety

$$\Xi_{\mathbb{C},x} \subset \mathbb{C}\mathbf{P}^{p-1}$$

has dimension $\ell - 1$ and degree σ. We shall say that the real characteristic variety

$$\Xi_x \subset \mathbf{P}^{p-1}$$

is hyperblic in case there is a real linear space $\mathbf{P}^{p-\ell-1}$ not meeting Ξ_x such that, for every $\xi \notin \mathbf{P}^{p-\ell-1}$, the linear space

$$\mathbf{P}^{p-\ell}(\xi) = \text{span}\{\mathbf{P}^{p-\ell-1}, \xi\}$$

meets Ξ_x in σ real points (with a suitable requirement concerning multiple roots).

When $\ell = p - 1$ we recover our previous definition in the hypersurface case. At the other extreme, when $\ell = 1$ then hyperbolicity means that Ξ_x should consist of σ distinct real points.

For us the importance of hyperbolicity lies in the following result (Yang [1]; also Bryant et al. [1]).

(23) Suppose that (I,J) is an involutive, hyperbolic Pfaffian differential system. Then the Cartan-Käehler theorem is true in the C^∞ category. The condition that (I,J) be hyperbolic is implied by the condition that its characteristic variety be hyperbolic in the following two cases:

$$\text{(24)} \quad \begin{aligned} &\text{(i)} \quad s_1' = \ldots = s_\ell' \neq 0, \ s_{\ell+1}' = \ldots = s_p' = 0, \ \ell < p \\ &\text{(ii)} \quad p = 3 \text{ with } s_1', s_2' \text{ arbitrary but } s_3' = 0. \end{aligned}$$

After all these preliminaries we now turn to the computation of the characteristic variety of the isometric embedding system for

$$\overline{X}^n \overset{x}{\rightarrow} X^n \subset \mathbf{E}^{n+r}. \tag{25}$$

Our goal is to prove the following

(26) THEOREM: i) The complex characteristic variety

$$\Xi_{\mathbb{C}} = \phi$$

if, and only if,

$$r \leq (n - 1)(n - 2)/2.$$

In general, for $r \geq (n - 1)(n - 2)/2$ we have

$$\dim \Xi_{\mathbb{C},x} = r - (n - 1)(n - 2)/2 - 1.$$

Note that when $r = n(n - 1)/2$ this gives

$$\dim \Xi_{\mathbb{C},x} = n - 2$$

in agreement with (15) and (5.9).

ii) If $n \geq 3$ and $r = n(n-1)/2$ then the real characteristic variety

$$\Xi \neq \phi.$$

If $n = 3$ and $\det(R_{ij}) \neq 0$, we may assume that it is hyperbolic.

This last statement means the following: Recall that the isometric embedding system is defined on $M \subset \mathcal{F}(\overline{X}) \times \mathcal{F}(E^{n+r}) \times \mathbb{R}^{rn(n+1)/2}$ where M is a smooth open subset of the solutions to the Gauss equations. Then if $n = 3$ and $\det(R_{ij}) \neq 0$ we may choose M to be non-empty and such that the isometric embedding system is hyperbolic there. Actually, a slightly weaker curvature condition is sufficient for this (cf. Bryant, Griffiths and Yang [1]). It is also proved in loc. cit. that the isometric embedding system is never hypobolic if $n \geq 4$. If $n = 4$ it is of principal type, but if $n \geq 6$ (and probably if $n \geq 5$) even this fails to be true. In any case, we may say that the pointwise structure of the (real and complex) characteristic variety is pretty well understood.

REMARK: What is not yet understood is the following: In general, for the isometric embedding system in the embedding dimension there is a stratification

$$\Xi = \Xi_o \supset \Xi_1 \supset \Xi_2 \supset \dots$$

where (loc. cit.)

$$\Xi_k = \text{singular locus of } \Xi_{k-1}$$
$$\operatorname{codim} \Xi_k = k^2.$$

For a manifold N, it is well known that there is a symplectic structure on T^*N leading to a Poisson bracket

$$\{,\}: C^\infty(T^*_{\mathbb{C}}N) \otimes C^\infty(T^*_{\mathbb{C}}N) \to C^\infty(T^*_{\mathbb{C}}N).$$

If $y^1, \dots, y^m$ are local coordinates on N, then points in $T^*_{\mathbb{C}}N$ are locally

$$\xi = \xi_i dy^i, \quad \xi^i \in \mathbb{C}.$$

The Poisson bracket of $f = f(y,\xi)$ and $g = g(y,\xi)$ is given by

$$\{f,g\}(y,\xi) = \frac{\partial f}{\partial y^i} \frac{\partial g}{\partial \xi_i} - \frac{\partial f}{\partial \xi_i} \frac{\partial g}{\partial y^i} .$$

We say that a subvariety $Z \subset T^*_{\mathbb{C}}N$ is involutive if its defining ideal is closed under $\{,\}$; i.e., if $f,g \in C^{\infty}(T^*_{\mathbb{C}}N)$ then

$$f\big|_N = g\big|_N = 0 \Rightarrow \{f,g\}\big|_N = 0.$$

At the other extreme, we may say that N is strongly non-involutive if the set of common zeroes of all the functions $\{f,g\}$ where $f\big|_N = g\big|_N = 0$ is empty. We would conjecture that, <u>for a general submanifold $X^n \subset \mathbf{E}^{n(n+1)/2}$, the subvarieties</u>

$$\tilde{\Xi}_k \subset T^*_{\mathbb{C}}X\backslash\{0\}$$

<u>are strongly non-involutive</u>. Here, $\tilde{\Xi}_k$ is the homogeneous cone lying over $\Xi_k \subset \mathbf{P}T^*_{\mathbb{C}}X$, and $\{0\} \subset T^*_{\mathbb{C}}X$ is the zero-section.

From (i) in (26) and (7) we have the following

(27) COROLLARY: The isometric embedding system for (25) is of finite type if

$$r \leq (n-1)(n-2)/2.$$

This is part (b) of Theorem (1.8). The rigidity result is proved by a further, non-linear, commutative algebra argument that we will not give here (cf. §III of Berger, Bryant, and Griffiths [2]).

From (ii) in (26) together with (23) we have the

(28) COROLLARY: If X is a 3-dimensional Riemannian manifold with $\det(R_{ij}) \neq 0$, then local C^{∞} isometric embeddings

$$x\colon \bar{X}^3 \to \mathbf{E}^6$$

exist.

Of course, it is conjectured that this result remains true for all n (assuming some non-degeneracy on the curvature), but even the corresponding linear P.D.E. is quite surprizingly not known to be locally solvable (cf. §IId of Bryant, Griffiths, and Yang [1] for further discussion). This corollary is part c) of Theorem (1.8) in the introduction.

We have now reduced everything to Theorem (26) above. In order to prove this theorem we must compute the symbol matrix for the isometric embedding system (I,J) given in (5.4). It is convenient to rewrite (5.4) as follows: $\{\underline{I}\}$ is generated by

$$(29)\quad \begin{aligned} &(\mathrm{i}) && \omega^i - \bar{\omega}^i = 0 \\ &(\mathrm{ii}) && \omega^\mu = 0 \\ &(\mathrm{iii}) && \omega^i_j - \bar{\omega}^i_j = 0 \\ &(\mathrm{iv}) && \theta^{\mu i} = \omega^\mu_i - h^\mu_{ij}\omega^j = 0, \qquad h^\mu_{ij} = h^\mu_{ji}. \end{aligned}$$

The structure equations are

$$(29)\quad \begin{aligned} &(\mathrm{v}) && d(\omega^i - \bar{\omega}^i) \equiv 0 && \mathrm{mod}\{\underline{I}\} \\ &(\mathrm{vi}) && d\omega^\mu \equiv 0 && \mathrm{mod}\{\underline{I}\} \\ &(\mathrm{vii}) && d(\omega^i_j - \bar{\omega}^i_j) \equiv 0 && \mathrm{mod}\{\underline{I}\} \\ &(\mathrm{viii}) && d\theta^{\mu i} \equiv -\pi^\mu_{ij} \wedge \omega^j && \mathrm{mod}\{\underline{I}\}. \end{aligned}$$

The symbol relations are

$$(30)\quad \begin{aligned} &(\mathrm{ix}) && \gamma(H,\pi) \equiv 0 \quad \mathrm{mod}\{\underline{J}\} \\ &(\mathrm{x}) && \pi^\mu_{ij} = \pi^\mu_{ji}. \end{aligned}$$

Our first observation is that this system is of the type of example (10). Thus (v), (vi) and (vii) may be ignored in computing the essential part of the symbol matrix.

To get warmed up we shall begin with the case $n = 2$, $r = 1$:

(31) Isometric embedding of a surface in $\mathbf{E}^3$.

In this case there are just two symbol relations in (31), one each in (ix) and (x). From (ix)

$$\gamma(H,\pi)_{1212} = h_{11}\pi_{22} + h_{22}\pi_{11} - 2h_{12}\pi_{12} = r_j^{1i}\pi_{ji},$$

where we have dropped the μ index here since its only value is now 3. From this we have

(32) $$r_2^{12} = h_{11},\ r_1^{11} = h_{22},\ r_2^{11} = -h_{12} = r_1^{12}.$$

From (x)

$$\pi_{12} - \pi_{21} = r_j^{2i}\pi_{ji},$$

we have

(33) $$r_2^{21} = 1 = -r_1^{22},\ r_1^{21} = 0 = r_2^{22}.$$

We shall denote a point in M by

$$H = (\bar{x},\bar{e},x,e_i,e_\mu,H) \in M \subseteqq \mathcal{F}(\bar{X}) \times \mathcal{F}(E^n) \times \mathcal{H},$$

because, as we will see, the symbol matrix depends only on the H coordinate. From (32) and (33)

(34)
$$\sigma_\xi(H) = (r^{ik}_j \xi_k) = \begin{vmatrix} h_{22}\xi_1 - h_{12}\xi_2 & h_{11}\xi_2 - h_{12}\xi_1 \\ -\xi_2 & \xi_1 \end{vmatrix}.$$

Consequently $\sigma_\xi(H)$ is not injective when

(35) $$\det \sigma_\xi(H) = h_{22}\xi_1\xi_1 + h_{11}\xi_2\xi_2 - 2h_{12}\xi_1\xi_2$$

is zero. The characteristic variety is the quadric

$$\Xi_H = \{[\xi] \in \mathbf{P}^2 : \det \sigma_\xi(H) = 0\}.$$

Observing that the discriminant of the quadratic form (35) is

$$h_{11}h_{22} - (h_{12})^2 = K,$$

where K is the Gaussian curvature of $\bar{X}$ (this is the Gauss equation (2.19) in this case), we infer that

(36.1) If $K > 0$, then $\Xi_{\mathbb{C},x}$ consists of two conjugate purely imaginary points while $\Xi_x = \phi$. In this case the isometric embedding system is elliptic (at the point $x \in M$).

(36.2) If $K < 0$, then $\Xi_x = \Xi_{\mathbb{C},x}$ consists of two distinct real points (corresponding to the asymptotic directions on X). In this case the isometric embedding system is hyperbolic.

(36.3) Finally, if $K = 0$ then $\Xi_x = \Xi_{\mathbb{C},x}$ consists of one real point counted twice.

Remark that the possibility $H = 0$ is ruled out by our assumption that M lie in the set of ordinary solutions to the Gauss equations.

To compute the characteristic variety in general, it will help to give the symbol matrix intrinsically, and for this some additional notation will be useful. We let

$$W \cong \mathbb{R}^r, \quad \text{with orthonormal basis} \quad w_\mu$$

be a Euclidean vector space representing a typical normal space, and

$$V^* \cong \mathbb{R}^n, \quad \text{with basis} \quad \omega^i$$

a vector space representing a typical cotangent space to $\overline{X}^n$. We set

$$S^qV^* = \mathrm{Sym}^q\ V^*, \quad q = 1,2,\dots .$$

For $\xi,\eta \in V^*$ we denote by $\xi\eta \in S^2V^*$ their symmetric product. In terms of components,

$$(\xi\eta)_{ij} = \frac{1}{2}(\xi_i\eta_j + \xi_j\eta_i).$$

As in Chapter 5 we denote by $K \subset \otimes^4V^*$ the space of <u>curvature-like tensors</u>

$$R = R_{ijk\ell}\omega^i \otimes \omega^j \otimes \omega^k \otimes \omega^\ell$$

satisfying the symmetries (listed in the proof of Theorem 5.9) of the Riemann curvature tensor. The dimension of K is given in (5.11).

The Gauss equations (2.20), or (2.21), may be considered as a quadratic mapping

$$\gamma : W \otimes V^* \otimes V^* \to \otimes^4V^* \tag{37}$$

which we denote by $\gamma(H,H)$, for $H \in W \otimes V^* \otimes V^*$, and which is defined by (2.21). The associated symmetric bilinear mapping

$$\gamma : S^2(W \otimes V^* \otimes V^*) \to \otimes^4V^* \tag{38}$$

is denoted $\gamma(H,G)$, for $H,G \in W \otimes V^* \otimes V^*$, (and is given below (5.7)). When restricted to $W \otimes S^2V^*$, or to $S^2(W \otimes S^2V^*)$, the image of γ lies in K.

For general n and r we see from (30) that there are $\dim K = n^2(n^2 - 1)/12$ symbol relations in (ix) and $rn(n - 1)/2$ symbol relations in (x). If we put

$$Q = n^2(n^2 - 1)/12 + rn(n - 1)/2,$$

then we describe the symbol matrix as a linear map

$$\sigma_\xi(H): W \otimes V^* \to \mathbb{R}^Q. \tag{39}$$

In terms of components

$$\sigma_\xi(H)^{\lambda i}_{\mu} = r^{\lambda j}_{\mu i}\xi_j, \quad \lambda = 1, \ldots, Q. \tag{40}$$

The first $n^2(n^2 - 1)/12$ rows of $\sigma_\xi(H)$ come from (ix) in (31). Here we denote the index λ by a quadruple $(ijk\ell)$. Then, from (ix),

$$\gamma(H,\pi)_{ijk\ell} = r^{(ijk\ell)m}_{\mu h}\pi^{\mu h}_{m} \qquad m,h = 1, \ldots, n \tag{41}$$

defines the right side.

The remaining $rn(n - 1)/2$ rows of $\sigma_\xi(H)$ come from (x) in (30). Here we denote the index λ by a triple (μij), $i < j$. Then, from (x),

$$\text{(42)} \qquad \pi^{\mu}_{ij} - \pi^{\mu}_{ji} = r^{(\mu ij)k}_{\sigma\ell}\pi^{\sigma\ell}_{k}$$

defines the right side.

Let

$$\text{(43)} \qquad \upsilon = t^{\mu}_{i}w_{\mu} \otimes \omega^{i} \in W \otimes V^{*}.$$

Using (41) and (42) we see that the mapping (39) is given by

$$\text{(44)} \qquad \begin{aligned} &\text{(i)} \quad (\sigma_{\xi}(H)\upsilon)^{(ijk\ell)} = \gamma(H,\upsilon\otimes\xi)_{ijk\ell} \\ &\text{(ii)} \quad (\sigma_{\xi}(H)\upsilon)^{(\mu ij)} = \xi_{i}t^{\mu}_{j} - \xi_{j}t^{\mu}_{i}. \end{aligned}$$

We can now prove

$$\text{(45)} \qquad \begin{aligned} &\{[\xi] \in \mathbf{P}V^{*}_{\mathbb{C}}\colon \sigma_{\xi}(H)\upsilon = 0 \quad \text{for some} \quad 0 \neq \upsilon \in W \otimes V^{*}\} \\ &= \{[\xi] \in \mathbf{P}V^{*}_{\mathbb{C}}\colon \gamma(H,\lambda\otimes\xi\xi) = 0 \quad \text{for some} \quad 0 \neq \lambda \in W\}. \end{aligned}$$

(46) **Remark** As H and $\lambda\otimes\xi\xi \in W \otimes S^{2}V^{*}$, where $\lambda \in W$, $\xi \in V^{*}$, we have, as noted below (38), that $\gamma(H,\lambda\otimes\xi\xi) \in K$. We may define a related "symbol mapping" by

(47)
$$\tilde{\sigma}_\xi(H) : W \to K.$$
$$\lambda \to \gamma(H, \lambda \otimes \xi\xi).$$

We shall comment more on this mapping below in (50).

To prove (45) we first observe that the second set is contained in the first, as we can take $\upsilon = \lambda \otimes \xi$. Suppose then that $\upsilon \neq 0$ is given by (43) and that

$$\sigma_\xi(H)\upsilon = 0.$$

Then (ii) in (44) says that

(48) $$\upsilon = \lambda \otimes \xi$$

for some $0 \neq \lambda \in W$. In fact,

$$\xi_i t_j^\mu - \xi_j t_i^\mu = 0 \quad \forall\, i, j$$

implies that for some $\lambda_{ij}^\mu \in \mathbb{C}$

$$(t_i^\mu, t_j^\mu) = \lambda_{ij}^\mu (\xi_i, \xi_j) \quad \text{(no sum)}.$$

As $\xi \neq 0$, we assume for the sake of argument that $\xi_1 \neq 0$. Then

$$(t_1^\mu, t_j^\mu) = \lambda_j^\mu(\xi_1, \xi_j) \quad \text{(no sum); i.e.,}$$

(49)

$$t_1^\mu = \lambda_j^\mu \xi_1 \quad \forall\ j.$$

Hence

$$\lambda_j^\mu = \lambda^\mu,$$

does not depend on j. By (49) then

$$t_j^\mu = \lambda^\mu \xi_j \quad \forall\ \mu, j,$$

which is (48) for $\lambda = \lambda^\mu w_\mu$.

We now obtain (45) from (i) of (44), as $\upsilon \otimes \xi = \lambda \otimes \xi \otimes \xi = \lambda \otimes \xi\xi$.

(50) **Remark**: The related symbol mapping (47) arises naturally if we examine the situation in P.D.E. theory.

Consider a linear P.D.E. 2nd order system

(51) $$r_\mu^{\lambda ij} \frac{\partial^2 z^\mu}{\partial x^i \partial x^j} = 0.$$

To compute its symbol matrix we may do either of two things:

(i) Define its symbol matrix as usual in P.D.E. theory by

$$\sigma_\xi = (r_\mu^{\lambda ij}\xi_i\xi_j). \tag{52}$$

(ii) Write (51) as the 1st order system

$$(53)\qquad \begin{aligned} r_\mu^{\lambda ij}\frac{\partial z_j^\mu}{\partial x^i} &= 0 \\ \frac{\partial z^\mu}{\partial x^i} &= z_i^\mu \end{aligned}$$

and then compute the symbol matrix of this big 1st order system. The point is that

> either of these two methods leads to the same characteristic variety.

This had certainly better be the case; we will omit the proof, which may be found in Bryant et al. [1].

Actually, (47) arises from the analogous result for the characteristic varieties of Pfaffian differential systems. Namely, suppose we consider (51) as the P.D.S.

$$(54)\qquad \begin{aligned} &\text{(i)}\quad \theta^\mu = dz^\mu - z_j^\mu dx^j \\ &\text{(ii)}\quad \theta_j^\mu = dz_j^\mu - z_{ji}^\mu dx^i, \quad z_{ij}^\mu = z_{ji}^\mu, \end{aligned}$$

defined on the subset M, of the space of variables $(x^i, z^\mu, z^\mu_i, z^\mu_{ij})$, given by

$$r^{\lambda ij}_\mu z^\mu_{ij} = 0.$$

The structure equations of (54) are

$$(55)\qquad \begin{aligned} d\theta^\mu &\equiv 0 \\ d\theta^\mu_j &\equiv -\pi^\mu_{ji} \wedge \omega^i , \end{aligned}$$

where $\pi^\mu_{ij} = dz^\mu_{ij}\big|_M$ and $\omega^i = dx^i\big|_M$, and the symbol relations are

$$(56)\qquad \begin{aligned} r^{\lambda ij}_\mu \pi^\mu_{ij} &= 0 \\ \pi^\mu_{ij} &= \pi^\mu_{ji} . \end{aligned}$$

If we compare the P.D.S (29), (30) with that of (54), (55), (56), we conclude that the symbol matrix of the former should be given by (52). For the former system, (52) translates into (47).

Equations (44) justify our earlier remark that the symbol matrix depends only on the H coordinate of a point of M. Furthermore, as shown in (45), the same is true of the characteristic variety which we write as Ξ_H. We now prove

$$\Xi_H = \{[\xi] \in \mathbf{P}V^*: \lambda^\mu H^\mu = \xi\eta \text{ for some } \lambda^\mu \in \mathbb{R},\ \eta \in V^*\}. \tag{57}$$

By (45), $[\xi] \in \Xi_H$ if, and only if,

$$\gamma(H, \lambda\otimes\xi\xi) = 0 \tag{58}$$

for some $0 \neq \lambda \in W$. Suppose that (58) holds. Write $\lambda = \lambda^\mu w_\mu$ and choose the basis for V^* so that $\xi = \omega^n$, which means that $\xi_i = \delta_{in}$. From the formula for $\gamma(H,\pi)$ below (5.7) we have

$$\begin{aligned} 2\gamma(H,\lambda\otimes\xi\xi) &= \\ &[h^\mu_{ik}\lambda^\mu\xi_j\xi_\ell + h^\mu_{j\ell}\lambda^\mu\xi_i\xi_k - \\ &\quad h^\mu_{i\ell}\lambda^\mu\xi_j\xi_k - h^\mu_{jk}\lambda^\mu\xi_i\xi_\ell]\omega^i\wedge\omega^j\otimes\omega^k\wedge\omega^\ell \\ &= \lambda^\mu h^\mu_{\alpha\beta}\omega^\alpha\wedge\omega^n\otimes\omega_\beta\wedge\omega^n, \end{aligned} \tag{59}$$

for $1 \leq \alpha,\beta \leq n-1$. Thus (58) is equivalent to

$$\lambda^\mu h^\mu_{\alpha\beta} = 0, \quad 1 \leq \alpha,\beta \leq n-1. \tag{60}$$

If we define the quadratic form $H^\mu \in SV^*$ by

(61) $$H^{\mu} = h^{\mu}_{ij}\omega^i\omega^j,$$

then (60) is equivalent to saying that

(62) $$\lambda^{\mu}H^{\mu} = \xi\eta$$

for some $\xi, \eta \in V^*$. Clearly (62) is independent of any special choice of basis. This proves (57).

To understand better the characteristic variety it is helpful to use a little algebraic geometry. For $\lambda = (\lambda^{\mu}) \in W_{\mathbb{C}}$ we set

$$H_{\lambda} = \lambda^{\mu}H^{\mu}$$

and view $|H_{\lambda}|_{[\lambda]\in \mathbf{P}W_{\mathbb{C}}}$ as a linear system of quadrics on the complex projective space $\mathbf{P}V^*_{\mathbb{C}}$ (linear systems of quadrics are almost the bread and butter of classical algebraic geometry). In $\mathbf{P}W_{\mathbb{C}}$ we have the determinantal loci

$$\Lambda_k = \{[\lambda] \in \mathbf{P}W_{\mathbb{C}} : \text{rank } H_{\lambda} \leq k\}.$$

For a general linear system of quadrics it may be shown by an elementary dimension count that (cf. Bryant, Griffiths, and Yang [1])

(63)
(i) $\operatorname{codim} \Lambda_k = (n-k)(n-k+1)/2$

(ii) $(\Lambda_k)_{sing} = \Lambda_{k-1}$.

(64) DEFINITION: We shall say that $H \in W \otimes S^2V^*$ is <u>non-degenerate</u> in case (63) is satisfied.

So, at long last we have defined the class of non-degenerate isometric embeddings with which these talks have been concerned: the Zariski open set $\mathcal{U}_{n,r}$ of the introduction consists of all H satisfying (63).

Motivated by (62) we define the incidence variety

$$\Sigma \subset \mathbf{P}W_{\mathbb{C}} \times \mathbf{P}V^*_{\mathbb{C}}$$

by the following standard algebro-geometric construction:

$$\Sigma = \{([\lambda],[\xi]) : H_\lambda = \xi\eta \quad \text{for some} \quad \eta \in V^*_{\mathbb{C}}\}.$$

The equation $H_\lambda = \xi\eta$ is equivalent to

$$\lambda^\mu h^\mu_{ij} = (\xi_i\eta_j + \xi_j\eta_i)/2;$$

it holds for some ξ and η if, and only if, rank $H_\lambda \leq 2$, (we are working over $\mathbb{C}$).

We therefore set $\Lambda_{\mathbb{C},H} = \Lambda_2$ and consider the diagram

(65)
$$\begin{array}{ccc} & \Sigma \subset \mathbf{P}W_{\mathbb{C}} \times \mathbf{P}V_{\mathbb{C}}^{*} & \\ {}^{\pi_1}\swarrow & & \searrow^{\pi_2} \\ \mathbf{P}W_{\mathbb{C}} \supset \Lambda_{\mathbb{C},H} & & \Xi_{\mathbb{C},H} \subset \mathbf{P}V_{\mathbb{C}}^{*} \end{array} .$$

The notation is justified, since by (63)

$$\Xi_{\mathbb{C},H} = \pi_2(\pi_1^{-1}\Lambda_{\mathbb{C},H}).$$

The situation is now as follows: (we omit proofs of the following intuitively plausible statements; cf. Bryant, Griffiths and Yang [1]):

(i) For a general point $[\xi] \in \Xi_{\mathbb{C},H}$ there are unique $[\lambda] \in \mathbf{P}W_{\mathbb{C}}$ and $\eta \in V_{\mathbb{C}}$ with

(66) $$H_{\lambda} = \xi\eta.$$

Thus, in the language of algebraic geometry the mapping

$$\Sigma \xrightarrow{\pi_2} \Xi_{\mathbb{C},H}$$

is <u>birational</u>. It may be shown (loc. cit.) that it is a resolution of singularities of $\Xi_{\mathbb{C},H}$.

(ii) Over a general point $[\lambda] \in \lambda_{\mathbb{C},H}$ there are two points of Σ, $([\lambda],[\xi])$ and $([\lambda],[\eta])$, where (66) holds. Thus, there is a rationally defined involution

$$\begin{aligned} &j:\Xi_{\mathbb{C},H} \to \Xi_{\mathbb{C},H} &&\text{given by} \\ &j([\xi]) = [\eta] &&\text{where (66) is satisfied.} \end{aligned}$$

(iii) The data of the characteristic sheaf is equivalent to giving the pair $(\Xi_{\mathbb{C},H}, j)$ (loc. cit.). Now it is not too difficult to see that the characteristic sheaf uniquely determines H up to $GL(W) \times GL(V)$ acting on $W \otimes S^2V^*$ (loc. cit.). From this together with one special fact in the case $n = 3$ we are led to the following (slightly corrected version of) a result of Tennenblatt [1]:

(67) Two isometric embeddings

$$x, x' : \overline{X}^3 \to \mathbf{E}^6$$

differ by a rigid motion if, and only if, at each $x \in \overline{X}$

$$(\Xi_{\mathbb{C},x}, j) = (\Xi'_{\mathbb{C},x}, j')$$

In other words, to have rigidity the characteristic varieties

$$\Xi_{\mathbb{C},x} \subset PT^*_{\mathbb{C},x}\overline{X}$$
$$\Xi'_{\mathbb{C},x} \subset PT^*_{\mathbb{C},x}\overline{X},$$

together with the involutions j, j', must coincide. The additional data beyond $\Xi_{\mathbb{C},H}$ of the involution j has proved crucial in understanding isometric embeddings, and it was this that pointed out the necessity of, in general, considering the characteristic sheaf.

Now we can at least outline the proof of Theorem (26). By what we have just said, using (63)

$$\begin{aligned}\dim \Xi_{\mathbb{C},H} &= \dim \Lambda_{\mathbb{C},H}\\ &= \dim \Lambda_2\\ &= r - 1 - (n-1)(n-2)/2,\end{aligned}$$

which implies i) in (26).

If $r = n(n-1)/2$ then

$$\begin{aligned}\dim \Xi_{\mathbb{C},H} &= \dim \Lambda_{\mathbb{C},H}\\ &= n - 2\end{aligned}$$

as expected. A standard algebro-geometric computation (cf. Bryant, Griffiths, and Yang [1]) gives that $\deg \Lambda_{\mathbb{C},H} = n$. When $n = 3$, thus $r = 3$, we may see this directly as follows: A general quadric in our linear system is

$$H_\lambda = \lambda^1 H^1 + \lambda^2 H^2 + \lambda^3 H^3 .$$

The condition $\operatorname{rank} H_\lambda \leq 2$ is not simply

$$\det(\lambda^1 H^1 + \lambda^2 H^2 + \lambda^3 H^3) = 0.$$

This is clearly a cubic curve in $\mathbf{P}W_{\mathbb{C}} = \mathbb{C}P^2$.

In general, as previously remarked, we may conclude that $\Xi_H \neq \phi$ for n odd. The proof that $\Xi_H \neq \phi$ for n even, $n \geq 4$ is based on a theorem of Adams, Lax, and Phillips [1]. Similar considerations appear already in the paper of Kaneda-Tannaka [1]. In Bryant, Griffiths, and Yang [1] it is also proved that $(\Xi_H)_{sing} \neq \phi$ for $n \geq 6$ (this should be true for $n \geq 5$), and by the discussion in loc. cit. we see that the linearized isometric embedding system does not fall into the class of known locally solvable linear P.D.E.'s. It is at this stage that the C^∞ local isometric embedding now stands.

Finally, the assertion, that, when $n = 3$ and $\det(R_{ij}) \neq 0$ we may assume that the characteristic cubic is hyperbolic (Fig. (2) below (22)), is proved by an explicit computation given also in loc. cit. □

We conclude this chapter with three remarks.

(68) REMARK: Given an isometric embedding

$$\overline{X}^n \overset{x}{\to} X \subset \mathbf{E}^{n(n+1)/2},$$

we view points $[\xi] \in \mathbf{P}T^*_x X$ as hyperplanes

$$\xi^\perp \subset T_x X \quad (\text{the annihilator of } \xi).$$

Let $\mathbf{E}^{n+1}_\lambda$ be the linear span of $T_x X$ and the normal vector $\lambda^\mu e_\mu$. By orthogonal linear projection we have a diagram

$$\begin{array}{c} X \subset \mathbf{E}^N \\ \downarrow \\ X_\lambda \subset \mathbf{E}^{n+1}_\lambda \end{array}$$

and H_λ is the 2nd fundamental form of X_λ at x. In fact, if (x, e_i, e_μ) is a Darboux frame for X, then $(x, e_i, \lambda^\mu e_\mu)$ is a Darboux frame for X_λ. Its 2nd fundamental form is

$$-\langle dx, d(\lambda^\mu e_\mu)\rangle = -\langle \omega^i e_i, \lambda^\mu(\omega^i_\mu e_i + \omega^\nu_\mu e_\nu)\rangle = \lambda^\mu h^\mu_{ij} \omega^i \omega^j = H_\lambda$$

Those hyperplanes $\xi^{\perp}$ corresponding to points $[\xi] \in \Xi_x$ are called asymptotic hyperplanes. They have the following geometric meaning: Suppose that (66) holds and write

$$(69) \qquad T_xX = (\xi^{\perp} \cap \eta^{\perp})_{\perp} \oplus (\xi^{\perp} \cap \eta^{\perp}) = T_{1\lambda} \oplus T_{2\lambda},$$

where $\perp$ as a subscript denotes orthogonal complement. Note that

$$(70) \qquad H_{\lambda}\Big|_{T_{2\lambda}} = 0.$$

We consider two cases.

(i) $[\xi] \neq [\eta]$. This is the general case. Then $T_{1\lambda} \cong \mathbb{R}^2$, $T_{2\lambda} \cong \mathbb{R}^{n-2}$, and by (70) X_{λ} is flat to 3rd order in the directions $T_{2\lambda}$; while in the directions $T_{1\lambda}$ it is to 3rd order a hyperbolic quadric. This just means that in a suitable coordinate system $(x^1,\dots,x^n,z)$ in $\mathbf{E}_{\lambda}^{n+1}$, X_{λ} is given by

$$(71) \qquad z = x^1x^2 + (\text{3rd order terms}).$$

In fact, it was noted in Chapter 2 that the 2nd fundamental form can be interpreted as meaning that for suitable coordinates $(y^1,\dots,y^n,z)$ in $\mathbf{E}_{\lambda}^{n+1}$, X_{λ} is given by

(72) $$z = \lambda^\mu h^\mu_{ij} y^i y^j + \text{(3rd order terms)}.$$

By (66) we get (71) from (72) by making a linear change of coordinates containing

$$x^1 = \xi_i y^i \ , \ x^2 = \eta_i y^i .$$

(ii) $[\xi] = [\eta]$. Then $T_{1\lambda} \cong \mathbb{R}$, $T_{2\lambda} \cong \mathbb{R}^{n-1}$. In a way similar to what was done in (i), one can show that now X_λ is locally given by

(73) $$z = (x^1)^2 + \text{(3rd order terms)}.$$

In particular, since $\Xi_x \neq \phi$ for $n \geq 3$ we conclude that the projections X_λ can never all be convex. In other words, for $n \geq 3$ there is no intrinsic curvature assumption that will guarantee the convexity of all projections to E^{n+1} of an $X^n \subset E^{n(n+1)/2}$. It is for this reason that the isometric embedding system for $\overline{X}^n \to E^{n(n+1)/2}$ always fails to be elliptic for $n \geq 3$ (cf. the comment below (18)).

(74) REMARK: Suppose we try to find a local isometric embedding $X^n \to X \subset E^{n(n+1)/2}$, where in a suitable

coordinate system $(y^1,\dots,y^n,z^{n+1},\dots,z^{n(n+1)/2}) = (y^i;z^\mu)$ the image X is given as a graph by

$$z^\mu = z^\mu(y).$$

Then, approximately, the isometric embedding equations imply that

$$(75)\qquad \sum_\mu \det \begin{vmatrix} \dfrac{\partial^2 z^\mu}{\partial y^i \partial y^k} & \dfrac{\partial^2 z^\mu}{\partial y^i \partial y^\ell} \\ \dfrac{\partial^2 z^\mu}{\partial y^j \partial y^k} & \dfrac{\partial^2 z^\mu}{\partial y^j \partial y^\ell} \end{vmatrix} = R_{ijk\ell}(y).$$

(The actual Gauss equations are a slight modification of (75), but the symbol properties are the same as those of (75)).

When $n = 2$ equation (75) is one Monge-Ampere equation for one function; these have been extensively studied.

When $n = 3$ the system (75) is 6 equations in 3 unknowns, which is "apparently" overdetermined. This is entirely consistent with (47), which implies that the symbol matrices are 1×1 in the case $n = 2$ and 6×3 in the case $n = 3$). (They are "overdetermined on the symbol level" whenever $n \geq 3$). On the other hand the characteristic variety has the same dimension as for a determined system, which is as it should be in the embedding dimension.

One of the early attempts at the C^∞ isometric embedding problem in the case $n = 3$ was to study systems (75). We feel that there is an obvious advantage in keeping everything intrinsic and, insofar as possible, in relegating all the analysis to the characteristic variety.

(76) REMARK: Given $R \in K$, the equations

$$\gamma(H,H) = R \tag{77}$$

have been the object of considerable study. Since γ is invariant under the group $O(r)$ of normal rotations we have

$$\dim \text{image } \gamma \leq rn(n+1)/2 - r(r-1)/2. \tag{78}$$

However, the equations (77) may have additional "hidden" symmetries. For example, when $n = 4$, $r = 2$ (corresponding to $X^4 \subset E^6$) it was proved in Berger, Bryant, and Griffiths [2] that

$$\dim K = 20$$

$$\text{right hand side of (78)} = 19$$

$$\dim (\text{image } \gamma) = 18.$$

Thus, there is a non-obvious extra fibre dimension of γ.

The Allendoerfer-Beez theorem mentioned in Chapter 1 is based on the fact that, if $r \leq [n/4]$ and $H, H' \in W \otimes S^2V^*$ are non-degenerate with

$$\gamma(H,H) = \gamma(H',H'),$$

then $H = A \cdot H'$ for some $A \in O(r)$. A recent algebraic study of the equations (77) is given in Steiner, Teufel, and Vilms [1].

CHAPTER 7

ISOMETRIC EMBEDDINGS OF SPACE FORMS

Two of the salient features of the theory of differential systems are i) that overdetermined systems are, at least formally, placed on an equal footing with determined systems (it is pretty clear that overdetermined systems will become increasingly important in geometry), and ii) that because of the intrinsic geometric character of the theory global questions should become accessible. In this section we will discuss two geometrically interesting overdetermined hyperbolic systems. Namely, we shall address the following

(1) PROBLEM: Let $(\overline{X}^n, \overline{ds}^2)$ be a Riemannian manifold of constant sectional curvature k. Determine all local isometric embeddings

(2) $$\overline{X} \overset{x}{\to} X \subset E^{n+r}$$

where r is minimal, and which are non-degenerate in the sense that their Gauss mapping is an immersion.

This problem was posed and solved by E. Cartan [3]. Since $(\overline{X},\overline{ds}^2)$ is real analytic the Cartan-Kahler theorem is applicable (when the initial data is analytic) so that the fact that the problem turns out to be hyperbolic is not so relevant. However, modifications of this problem lead to proper C^∞ questions, and for this reason we view its hyperbolicity as important. We shall restrict our attention to the case $k \leq 0$.

Case 1: $k = 0$. We begin by studying the Gauss equations of

$$X^n \subset \mathbf{E}^{n+r} \tag{3}$$

where X is flat. For this, let W be a real Euclidean vector space of dimension r with inner product $\cdot$, and let V be a real n-dimensional vector space (without a given inner product). Elements of $W \otimes S^2V^*$ will be called 2nd fundamental forms. A 2nd fundamental form

$$H \in W \otimes S^2V^*$$

is said to be non-degenerate in case H involves all the variables of V^*; i.e.,

$$H \not\in W \otimes S^2U$$

for any proper subspace $U \subset V^*$. (Note: this use of non-degenerate, which is the standard one from linear algebra when $\dim W = 1$, is special to Section 7 of these talks - it agrees with Cartan's terminology but differs from our previous terminology.) A submanifold (3) is non-degenerate if its 2nd fundamental form $II(x) \in N_x \otimes S^2T_x^*$ is non-degenerate at each $x \in X$. This is equivalent to the Gauss mapping

$$X \to G(n, n + r)$$
$$x \to T_xX$$

having maximal rank.

We shall restrict our attention to non-degenerate H's.

DEFINITION (CARTAN): $H \in W \otimes S^2V^*$ is <u>exteriorly orthogonal</u> if

$$(4) \qquad \gamma(H, H) = 0,$$

where γ is defined in (2.21). Clearly, these are just the Gauss equations for a flat submanifold (3). The main algebraic fact, which we shall not prove (cf. Cartan [3], Spivak [1], or Moore [1] is the

(5) THEOREM (Cartan): If $H \in W \otimes S^2V^*$ is non-degenerate and exteriorly orthogonal, then $r \geq n$. If $r = n$ then there is an orthonormal basis $w_1, \dots, w_n$ for W and basis $\varphi^1, \dots, \varphi^n$ for V^* such that

(6) $$H = \sum_i w_i \otimes (\varphi^i)^2.$$

Conversely, every H of the form (6) is a exteriorly orthogonal.

The set of H of the form (6) is a quotient of $O(n) \times GL(n,\mathbb{R})$ by a finite subgroup (permutations plus changes of sign of the φ^i); we let

$$\mathcal{V}_n \subset W \otimes S^2V^*$$

be the (smooth) variety of 2nd fundamental forms (6).

Let us now set up, as in Chapter 5, the P.D.S. giving the non-degenerate isometric embeddings (2) in E^{2n} where $(\overline{X}^n, \overline{ds}^2)$ is flat. This is the P.D.S. (5.4)-(5.5), or rewritten (6.29)-(6.30), on

$$M = \mathcal{F}(\overline{X}) \times \mathcal{F}(E^{2n}) \times \mathcal{V}_n.$$

We shall compute the characteristic variety of this system. For this, remark that, just as in Chapter 6, the characteristic variety at a point of M depends only on the $H \in \mathscr{V}_n$ component; we denote this variety by

$$\Xi_H \subset \mathbf{P}V^*.$$

Suppose that H is given by (6).

(7) PROPOSITION: Ξ_H consists of the $n(n-1)/2$ lines joining two of the points

$$[\varphi^1], \ldots, [\varphi^n] \in \mathbf{P}V^*.$$

The complex characteristic variety consists of the corresponding complex lines.

PROOF: If $[\xi] \in \Xi_H$, then by (6.57) we must have

$$\lambda \cdot H = \xi\eta \tag{8}$$

for some $\lambda \in W$ and $0 \neq \eta \in V^*$. Conversely, if (8) holds then $[\xi] \in \Xi_H$. As H is given by (6), if we set $\lambda = \lambda^i w_i$, then equation (8) is

(9) $$\sum_i \lambda^i (\varphi^i)^2 = \xi\eta.$$

The quadratic form on the left hand side has rank ≤ 2 if, and only if, at most two of the λ_i are non-zero, say $\lambda_i = 0$ unless $i = 1,2$. Then necessarily

$$\xi = \mu_1\varphi^1 + \mu_2\varphi^2,$$

and the proposition follows. □

Here is the picture in the case $n = 3$

Figure 1

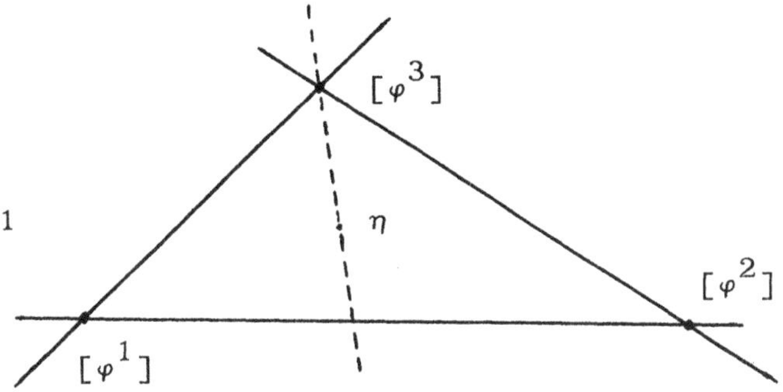

From this result together with (6.15) and (6.23) we may draw the following conclusion:

(10) If the isometric embedding system on the above M is involutive, then the local isometric embeddings (2) depend on $n(n - 1)/2$ arbitrary functions of two variables. Moreover, the system is hyperbolic.

Remark that in the above figure, the (real) lines through η meet Ξ_H in three points that may not be distinct; however, the system is still hyperbolic due to the lack of "branching" of the roots (the same is true for all n).

Involutivity of the above system was not established in Chapter 6 for the cases $n \geq 3$ because, for these values of n, $\mathcal{V}_n$ is not contained in $\mathcal{H}$, the ordinary second fundamental forms. It is true that the present isometric embedding system is involutive, but the proof of this is clumsy.

For the purposes of establishing involutivity we will set up a slightly different P.D.S. with essentially the same integrals and having the same characteristic variety. Let **F** denote the bundle of frames $f = (x;e_i;e_{n+j})$ of $\mathbf{E}^{2n}$ that satisfy

$$(11)\qquad \begin{aligned} &(i) \quad e_i \cdot e_{n+j} = 0 \\ &(ii) \quad e_{n+i} \cdot e_{n+j} = \delta^i_j . \end{aligned}$$

The motivation here is that the φ^i in (6) may not be orthonormal. On **F** we set

$$g_{ij} = e_i \cdot e_j$$
$$G = (g_{ij}) = {}^tG,$$

and define 1-forms ω^a, ω^a_b on **F** by

(12)
$$\text{(i)} \quad dx = \omega^a e_a$$
$$\text{(ii)} \quad de_a = \omega^b_a e_b \quad (a,b = 1,\dots,n + r = 2n).$$

Upon differentiating (11) and substituting in (ii) of (12), we observe that

(13)
$$\omega^{n+i}_{n+j} + \omega^{n+j}_{n+i} = 0$$
$$\omega^{n+j}_i + g_{ik}\omega^k_{n+j} = 0.$$

Moreover, the 1-forms

$$\omega^i,\ \omega^i_j,\ \omega^{n+i},\ \omega^{n+j}_i,\ \omega^{n+j}_{n+i}\ (i < j)$$

give a coframing of **F**. Exterior differentiation of (12) gives

(14)
$$d\omega^a = \omega^b \wedge \omega^a_b$$
$$d\omega^b_a = \omega^c_a \wedge \omega^b_c$$

We now let (I,J) be the Pfaffian differential system on $\mathbf{F}$ given by

$$(15)\qquad \begin{aligned} &(i) && \omega^{n+i} = 0 \\ &(ii) && \omega^{n+i}_j - \delta^i_j \omega^j = 0 \qquad \text{(no summation)} \\ &(iii) && \omega^1 \wedge \ldots \wedge \omega^n \neq 0. \end{aligned}$$

According to example (3.3) and Theorem 5 above, the local integral manifolds are in a one-to-one correspondence with non-degenerate local isometric embeddings (2) when $r = n$ and $(\overline{X}, \overline{ds}^2)$ is flat. Indeed, the problem has now become to find all flat submanifolds $X^n \subset E^{2n}$, and these are the integral manifolds of (15).

In fact, one needs to observe the following. Suppose that we have an isometric embedding (2), and suppose that (x, e_i, e_μ) is a frame along x satisfying

i) e_i are tangent to X, but not necessarily orthonormal;

ii) e_μ are normal to X and orthonormal.

Then (12) holds with (i) of (12) changed to

$$dx = \omega^i e_i,$$

and (13) holds where

$$g_{ij} = \langle e_i, e_j \rangle.$$

Furthermore (14) holds, from which we get (since $\omega^{\mu} = 0$)

$$\omega_i^{\mu} = h_{ij}^{\mu}\omega^j, \quad h_{ij}^{\mu} = h_{ji}^{\mu}.$$

The point to be made here is that the second fundamental form of X is

$$H = h_{ij}^{\mu} e_{\mu} \otimes \omega^i \omega^j,$$

which follows from

$$H = -e_{\mu} \otimes \langle dx, de_{\mu} \rangle = -e_{\mu} \otimes \langle \omega^i e_i, \omega_{\mu}^j e_j + \omega_{\mu}^{\upsilon} e_{\upsilon} \rangle$$

$$= -e_{\mu} \otimes \omega^i g_{ij} \omega_{\mu}^j = e_{\mu} \otimes \omega^i \omega_i^{\mu},$$

by (13).

(16) PROPOSITION: The Pfaffian differential system (15) is involutive and hyperbolic. Moreover,

$$s_1' = n^2, \quad s_2' = n(n - 1)/2, \quad s_3' = \ldots = s_n' = 0.$$

Thus the non-degenerate flat submanifolds

$$X^n \subset E^{2n}$$

locally depend on $n(n-1)/2$ functions of two variables.

PROOF: During this proof we will not use the summation convention. The Maurer-Cartan equations (14) give for the structure equations of the P.D.S. (15)

$$(17)\qquad \begin{aligned} &d\omega^{n+i} \equiv 0 \\ &d(\omega_j^{n+i} - \delta_j^i\omega^j) \equiv \sum_k (\delta_j^i\omega_k^j + \delta_k^i\omega_j^k - \delta_k^j\omega_{n+k}^{n+i})\wedge\omega^k, \end{aligned}$$

where $\equiv$ denotes congruence modulo $\{\underline{I}\}$. Write the 2nd equation as

$$(18)\qquad d(\omega_j^{n+i} - \delta_j^i\omega^j) \equiv -\sum_k \pi_{jk}^i\wedge\omega^k,$$

so that the tableau matrix is

$$\pi = \begin{vmatrix} 0 \\ \cdots\cdots \\ \pi^1_{11} \cdots \pi^1_{1n} \\ \vdots \qquad \vdots \\ \pi^1_{n1} \cdots \pi^1_{nn} \\ \vdots \qquad \vdots \\ \pi^n_{11} \cdots \pi^n_{1n} \\ \vdots \qquad \vdots \\ \pi^n_{n1} \cdots \pi^n_{nn} \end{vmatrix}$$

As always we may ignore the blocks of zeroes. Then the kth column is an $(n \times n)$-matrix-valued 1-form $\pi_k = (\pi^i_{jk})_{1 \le i,j \le n}$.

At this juncture an interesting point arises. Namely, by definition $s'_1 + \ldots + s'_k$ is for $1 \le k \le n-1$ the number of independent 1-forms modulo $\{\underline{J}\}$ in the first k columns of π for a generic flag, and it is pretty clear that the flag given by $\{\omega^1\}$, $\{\omega^1, \omega^2\}, \ldots$ is non-generic (in fact, ω^i corresponds to the ith vertex of the characteristic variety).

A generic flag is obtained from a generic basis $\tilde{\omega}^i$ of J/I, which is related to ω^i by

$$\omega^i = \sum_j t^i_j \tilde{\omega}^j ,$$

where $(t^i_j) \in GL(n;\mathbb{R})$ is generic. Then the right side of (18) becomes

$$-\sum_k \pi^i_{jk} \wedge \sum_\ell t^k_\ell \tilde{\omega}^\ell = -\sum_\ell \tilde{\pi}^i_{j\ell} \wedge \tilde{\omega}^\ell,$$

where

$$\tilde{\pi}^i_{j\ell} = \sum_k t^k_\ell \pi^i_{jk}.$$

Hence

$$\tilde{\pi}_\ell = \sum_k t^k_\ell \pi_k.$$

Thus, e.g., s_1' is the number of independent 1-forms in the $n \times n$ matrix

$$\pi(\lambda) = \pi_k \lambda^k$$

for $\lambda^k = t^k_1$, $\lambda = (\lambda^1, \ldots, \lambda^n)$. We have from (17), (18)

$$\pi(\lambda)^j_j = 2\omega^j_j \lambda^j + \sum_{k \neq j} \omega^j_k \lambda^k$$

$$\pi(\lambda)^i_j = \omega^i_j \lambda^i - \omega^{n+i}_{n+j} \lambda^j \,, \quad i \neq j.$$

We claim that, if all $\lambda^i \neq 0$, then these n^2 1-forms are linearly independent; in fact we will show that they are linearly independent modulo $S = \operatorname{span}\left\{\omega^{n+i}_{n+j}\right\}$. Denote by $\overline{\varphi}$ the 1-form φ considered modulo S. The 1-forms $\overline{\omega}^i_j \lambda^i$ $(i \neq j)$ are linearly independent, and then the 1-forms

$$2\bar{\omega}^j_j\lambda^j + \sum_{k\neq j} \bar{\omega}^j_k\lambda^j(\lambda^k/\lambda^j)$$

are linearly independent from these. Thus

$$s_1' = n^2.$$

Next, we shall show that if

$$(*) \qquad \begin{aligned} &\lambda = (\lambda^1, \ldots, \lambda^n), \quad \lambda^i \neq 0 \\ &\eta = (\eta^1, \ldots, \eta^n) \\ &\lambda^i\eta^j - \lambda^j\eta^i \neq 0, \quad \text{for all} \quad i \neq j \end{aligned}$$

(i.e. $\eta^i = t^i_2$), then from among the $2n^2$ 1-forms

$$(19) \qquad \pi(\lambda)^i_j, \ \pi(\eta)^i_j$$

exactly $n^2 + n(n-1)/2$ are linearly independent.
For $i \neq j$

$$\pi(\lambda)^i_j\eta^i - \pi(\eta)^i_j\lambda^i = \omega^{n+i}_{n+j}(\lambda^i\eta^j - \lambda^j\eta^i).$$

Choosing λ and η to satisfy (*), (e.g., take $\lambda^j = 1$ and $\eta^j = j$, for all j), we see that each ω^{n+i}_{n+j} $(i < j)$ is a linear combination of 1-forms in (19). Thus the span of the forms in (19) contains the $n^2 + n(n-1)/2$ independent forms ω^i_j, ω^{n+i}_{n+j} $(i < j)$, and hence

$$s_2' = n(n - 1)/2.$$

Furthermore $s_3' = \ldots = s_n' = 0$ because we have obtained from the first two columns of the generic reduced tableau matrix all the 1-forms which are non-zero modulo $\{\underline{J}\}$.

REMARK: For future reference we characterize the vectors λ, η satisfying (*). In an n-dimensional vector space V with basis $v_1, \ldots, v_n$ and dual basis $\bar{\omega}^1, \ldots, \bar{\omega}^n$ of V^* consider the 2-planes E spanned by the linearly independent vectors $\lambda = \lambda^i v_i$ and $\eta = \eta^i v_i$. Let $G_2(V)$ denote the Grassmannian of 2-planes in V. Then

$$\lambda^i \eta^j - \lambda^j \eta^i = \langle \bar{\omega}^i \wedge \bar{\omega}^j, \lambda \wedge \eta \rangle,$$

and thus the set of vector λ, η satisfying (*) is equivalent to

$$\{E \in G_2(V) : \bar{\omega}^i \wedge \bar{\omega}^j \big|_E \neq 0, \quad \text{for all} \quad i < j\}.$$

To establish involutivity of (15) we must show that the integral elements of (15) depend on

$$s_1' + 2s_2' = n^2 + n(n-1)$$

parameters. Let A^i_j $(i \neq j)$ and B^i_j be $n(n-1) + n^2$ arbitrary real numbers and consider the n-plane

$$\begin{aligned}
&\omega^{n+i} = \omega^{n+i}_j - \delta^i_j\omega^j = 0 \\
&\omega^{n+i}_{n+j} - (A^i_j\omega^i - A^j_i\omega^j) = 0 \qquad i \neq j \\
&\omega^j_j - \sum_k B^j_k\omega^k = 0 \\
&\omega^i_j - (2B^i_j\omega^i - A^i_j\omega^j) = 0 \qquad i \neq j.
\end{aligned}$$

It follows from (17) that this is an integral element, and, by Cartan's Test, we have established involutivity of (15).

It remains to show that the P.D.S. (15) has the same characteristic variety (7) as that given by the usual isometric embedding system (6.29)-(6.30). With a little work this may be done by direct computation - i.e., by computing the symbol relations of the forms

$$\pi^i_{jk} = -(\delta^i_j\omega^j_k + \delta^i_k\omega^k_j - \delta^j_k\omega^{n+i}_{n+k})$$

appearing in (17), and from these symbol relations computing the characteristic variety as the locus where the symbol matrix σ_ξ fails to be injective.

However, using the involutivity of (15), there is an alternate approach based on a general result given in

Bryant et al. [1]. As this general result is one of the deeper results on characteristic varieties of involutive systems, and as it will also apply to the computation of the characteristic variety of the hyperbolic space form embedding system (32) below, we shall state it here and show how it may be applied to the problem at hand.
Thus, we give the following

DIGRESSION: Let (I,J) be a quasi-linear Pfaffian differential system on a manifold M given by (cf. (6.2))

$$\begin{aligned} &\theta^{\alpha} = 0 && \alpha = 1,\dots,s \\ &d\theta^{\alpha} \equiv -\pi^{\alpha}_{\rho}\wedge\omega^{\rho} \mod\{\theta\} && \rho = 1,\dots,p \\ &\omega^1\wedge\dots\wedge\omega^p \neq 0. && \end{aligned}$$

The summation convention is back in effect, and we shall use the notation introduced at the beginning of Chapter 6. Thus, for any $x \in M$, $V^*_x = J_x/I_x$, where $I_x \subset J_x \subset T^*_xM$. Using general theorems about dual spaces, we obtain

$$V_x = I^{\perp}_x/J^{\perp}_x,$$

where $I^{\perp}_x$ and $J^{\perp}_x$ are the annihilators of I_x and J_x in T_xM. For example,

$$I_x^{\perp} = \{v \in T_xM : \theta(v) = 0 \ \forall \ \theta \in I_x\}.$$

Thus, the elements of V_x are tangent vectors in T_xM modulo vectors annihilated by J_x.

For a 1-form $\omega \in J_x$ we denote its equivalence class in V_x^* by $\bar{\omega}$. Thus $\bar{\omega}^1, \ldots, \bar{\omega}^p$ is a basis of V_x^*, and we let $\bar{v}_1, \ldots, \bar{v}_p$ denote the dual basis of V_x, where $\bar{v}_\rho$ denotes the equivalence class of $v_\rho \in I_x^{\perp}$.

If the symbol relations for our system are (cf. (6.2) iii)

$$r_\alpha^{\lambda\rho}(x)\pi_\rho^\alpha(x) \equiv 0 \quad \mathrm{mod}\{J_x\}, \quad \lambda = 1, \ldots, Q,$$

then for each $x \in M$ the characteristic variety

$$\Xi_x \subset \mathbf{P}V^*$$

is defined by (cf. (6.4))

$$\Xi_x = \{[\xi] \in \mathbf{P}V^* : r_\alpha^{\lambda\rho}(x)\xi_\rho\eta^\alpha = 0 \quad \text{for some} \quad \eta \neq 0\}.$$

We shall work over a fixed point $x \in M$, and shall drop refence to it.

Suppose now that (I, J) is involutive and has character ℓ, (see defintion above (4.26)) - i.e.,

$$s'_\ell \neq 0,\ s'_{\ell+1} = \ldots = s'_p = 0.$$

For example, (15) has character two and Cartan integer $s'_2 = n(n-1)/2$. According to the proof of the Cartan-Kähler theorem, in the real analytic case we may find local integral manifolds of (I, J) by posing an initial value problem along a "general" ℓ-dimensional integral submanifold $N^\ell \subset M$, where the initial data depends on s'_ℓ arbitrary functions given on N.

To clarify the meaning of "general", we consider admissible ℓ-dimensional integral elements $E \subset T_xM$. Such an ℓ-plane must be contained in $I_x^\perp$, and to be admissible requires that

$$\omega^1 \wedge \ldots \wedge \omega^p \big|_E \neq 0,$$

which means that $E \cap J_x^\perp = (0)$. Hence the projection of E, which we continue to denote E, under

$$I_x^\perp \to I_x^\perp / J_x^\perp = V$$

is an ℓ-dimensional subspace of V. In this way, we may consider the admissible ℓ-dimensional integral elements at $x \in M$ as being contained in the Grassmannian $G_\ell(V)$ of ℓ-planes in V.

We shall define the Cartan characteristic variety

$\Lambda \subset G_\ell(V)$, and then the initial manifold N will be "general" in case it is non-characteristic in the sense that $T_xN \notin \Lambda$ for all $x \in N$. Thus, the Cartan characteristic variety addresses the P.D.E. meaning of the word "characteristic", whereas the usual characteristic variety is defined by properties of the symbol map. The subtlety is that, in what might be called the overdetermined case when $\ell < p-1$, the Cartan characteristic variety $\Lambda \subset G_\ell(V)$ and the usual characteristic variety $\Xi \subset \mathbf{P}V^* = G_{p-1}(V)$ live in different places.

To define Λ we consider ℓ-planes $E \subset V$ with basis $\{e_t\}$, $1 \leq t \leq \ell$. Write $e_t = t_t^\rho \bar{v}_\rho$ and let $\pi(E)$ be the collection of 1-forms $\{t_t^\rho \pi_\rho^\alpha\}$ considered modulo J. In other words, $\pi(E)$ consists of the 1-forms in the first ℓ columns of the tableau matrix set up with respect to the basis $\tilde{\omega}^\rho$ of V^*, where $\bar{\omega}^\rho = t_\sigma^\rho \tilde{\omega}^\sigma$, where in turn the $p \times p$ matrix (t_σ^ρ) has for its first ℓ columns the above t_t^ρ. (Cf. the proof of Proposition (16))

DEFINITION: The <u>Cartan characteristic variety</u> $\Lambda \subset G_\ell(V)$ is defined by

$$\Lambda = \{E \in G_\ell(V): \dim \pi(E) < s_1' + \ldots + s_\ell'\}.$$

In the obvious manner we also define $\Lambda_{\mathbb{C}} \subset G_{\ell}(V_{\mathbb{C}})$. Clearly, Λ is a proper algebraic subvariety of $G_{\ell}(V)$.

We now identify points $[\xi] \in \mathbf{P}V^*$ with hyperplanes $\xi^{\perp} = \{v \in V: \langle \xi, v \rangle = 0\}$ in V; thus

$$\mathbf{P}V^* = G_{p-1}(V).$$

The relationship between Cartan characteristic varieties and usual characteristic varieties is given by the following

THEOREM: Suppose that (I,J) is an involutive quasi-linear P.D.S. Then

$$\Xi_{\mathbb{C}} = \{[\xi]: \text{ every } E \subset \xi^{\perp} \text{ belongs to } \Lambda_{\mathbb{C}}\}$$

$$\Lambda_{\mathbb{C}} = \{E: E \subset \xi^{\perp} \text{ for some } [\xi] \in \Xi_{\mathbb{C}}\}.$$

REMARKS: This result, which requires both involutivity and the use of complex characteristic varieties, is one of the deepest properties of characteristic varieties. As shown by the Cauchy-Riemann equations (cf. Bryant et al., [1]) we may well have $\Xi = \phi$ but $\Lambda \neq \phi$.

We now complete the proof of (16) in which it remains to show that (15) is hyperbolic. To do this we apply the above result to compute the characteristic variety $\Xi \subset \mathbf{P}^{n-1}$ of (15). This requires that we compute the Cartan characteristic variety $\Lambda \subset G_2(V)$ of (15).

From the Remark made above, at the end of the computation of the reduced characters of (15), we conclude that the complex Cartan characteristic variety of (15) is

$$\Lambda_{\mathbb{C}} = \{E \in G_2(V_{\mathbb{C}}) : \overline{\omega}^i \wedge \overline{\omega}^j \big|_E = 0 \quad \text{for some} \quad i < j\}.$$

Suppose that $[\xi] \in P(V_{\mathbb{C}}^*)$, and that ξ has the form

$$\xi = a\overline{\omega}^i + b\overline{\omega}^j,$$

for some $i < j$, $a, b \in \mathbb{C}$. Then

$$(a\overline{\omega}^i + b\overline{\omega}^j)\big|_{\xi^\perp} = 0,$$

which means that $\overline{\omega}^i\big|_{\xi^\perp}$ and $\overline{\omega}^j\big|_{\xi^\perp}$ are linearly dependent and thus

$$\overline{\omega}^i\big|_{\xi^\perp} \wedge \overline{\omega}^j\big|_{\xi^\perp} = 0.$$

In particular, for any 2-plane $E \subset \xi^\perp$,

$$(\bar{\omega}^i \wedge \bar{\omega}^j)\big|_E = 0;$$

i.e., $E \in \Lambda_{\mathbb{C}}$. Hence $[\xi] \in \Xi_{\mathbb{C}}$.

On the other hand, suppose that $[\xi] \in P(V^*_{\mathbb{C}})$ has the form $\xi = \xi_i \bar{\omega}^i$, where for at least three values of i we have $\xi_i \neq 0$. Now the only linear relation among $\bar{\omega}^i\big|_{\xi^\perp}$, $i = 1,\ldots,n$ is $\xi_i \bar{\omega}^i\big|_{\xi^\perp} = 0$. Thus for any $i < j$, $\bar{\omega}^i\big|_{\xi^\perp}$ and $\bar{\omega}^j\big|_{\xi^\perp}$ must be linearly independent; i.e., $(\bar{\omega}^i \wedge \bar{\omega}^j)\big|_{\xi^\perp}$ is non-zero in $\Lambda_2(\xi^\perp)^*$. Hence

$$\{\ker(\bar{\omega}^i \wedge \bar{\omega}^j)\big|_{\xi^\perp} : 1 \leq i < j \leq n\}$$

is a set of at most $n(n-1)/2$ distinct hyperplanes in $\Lambda_2(\xi^\perp)^*$. There exists, then, a decomposable element $\lambda \wedge \eta \in \Lambda_2(\xi^\perp)^*$, where $\lambda, \eta \in \xi^\perp$, not in one of these hyperplanes. For $E = \{\lambda, \eta\} \in G_2(\xi^\perp)$ we have $(\bar{\omega}^i \wedge \bar{\omega}^j)\big|_E \neq 0$ for all $i < j$ - i.e., $E \notin \Lambda_{\mathbb{C}}$ - and hence $[\xi] \notin \Xi_{\mathbb{C}}$.

In conclusion, we have shown that $\Xi_{\mathbb{C}}$ consists of the $n(n-1)/2$ complex lines joining the points $[\bar{\omega}^i] \in P(V^*_{\mathbb{C}})$, $i = 1,\ldots,n$. This agrees with Proposition (7) and, finally, completes the proof of Proposition (16). □

CASE 2: $k = -1$. We now let $(\overline{X}, \overline{ds}^2)$ be a Riemannian manifold of constant sectional curvature -1. (The summation convention is back in force). On $\mathcal{F}(\overline{X})$ the structure equations (2.8), (2.9) are

$$\begin{aligned} d\overline{\omega}^i &= -\overline{\omega}^i_j \wedge \overline{\omega}^j \\ d\overline{\omega}^i_j + \overline{\omega}^i_k \wedge \overline{\omega}^k_j &= -\overline{\omega}^i \wedge \overline{\omega}^j. \end{aligned} \tag{20}$$

The 2nd equation may be written as

$$R = -\gamma(\overline{ds}^2, \overline{ds}^2) \tag{21}$$

where γ, given by (2.21), is the ubiquitous Gauss equation mapping acting on $\overline{ds}^2 \in \mathbb{R} \otimes S^2(T^*_x\overline{X})$. Using Theorem (5) we shall show that:

(22) For any local isometric embedding (2), we have

$$r \geq n - 1.$$

PROOF: Suppose such an embedding exists. For $x \in X$ set $W = N_xX \cong \mathbb{R}^r$, $V = T_xX \cong \mathbb{R}^n$, and let $H \in W \otimes S^2V^*$ be the 2nd fundamental form $II(x)$. For

$$\hat{W} = W \oplus \mathbb{R}$$

$$\hat{H} = H \oplus ds^2 \in W \otimes S^2V^* \oplus \mathbb{R} \otimes S^2V^* = \hat{W} \otimes S^2V^*$$

the Gauss equations plus (21) give

$$\gamma(\hat{H},\hat{H}) = \gamma(H,H) + \gamma(ds^2,ds^2) = 0.$$

Moreover, due to the ds^2 factor it is clear that $\hat{H}$ is non-degenerate (in the sense of this section). Theorem (5) then gives

$$r + 1 = \dim \hat{W} \geq n. \qquad \square$$

We shall now study local isometric embeddings

(23) $$\overline{X}^n \to X^n \subset E^{2n-1}$$

in the minimal possible codimension $n - 1$. Concerning these the main result is the

(24) THEOREM: (Due essentially to Cartan [3]): The isometric embedding system for (23) is involutive and hyperbolic with Cartan characters

$$s_1' = n(n-1), \quad s_2' = \ldots = s_n' = 0.$$

COROLLARY: Local isometric embeddings (23) exist and depend on $n(n-1)$ functions of one variable.

Put geometrically, the initial value problem for (23) is posed along a curve in E^{2n-1}.

Before proving this theorem we shall first see what such embeddings must look like, and then shall work backwards. The summation convention will no longer be used.

We fix a point $x \in X$, keep the above notation, and use (6) to infer that

$$H \oplus ds^2 = \sum_1 \hat{w}_i \otimes (\varphi^i)^2 \tag{25}$$

for some orthonormal basis $\hat{w}_1, \ldots, \hat{w}_n$ of $\hat{W} = W \oplus \mathbb{R}$ and some basis (not necessarily orthonormal) $\varphi^1, \ldots, \varphi^n$ of V^*. Setting

$$\hat{w}_i = w_i \oplus r_i, \quad w_i \in W, \ r_i \in \mathbb{R},$$

we get from (25)

$$ds^2 = \sum_i r_i (\varphi^i)^2.$$

Thus $r_i > 0$ and we define

$$\omega^i = \sqrt{r_i}\,\varphi^i$$

$$b_i = w_i / r_i .$$

As the $\hat{w}_i$ are orthonormal, we have

$$\delta_{ij} = \hat{w}_i \cdot \hat{w}_j = r_i r_j (b_i \cdot b_j + 1).$$

Hence

(26) $$b_i \cdot b_j = -1, \quad \text{if} \quad i \neq j.$$

Furthermore,

(27)
(i) $ds^2 = \sum_i (\omega^i)^2$
(i.e., $\omega^1, \ldots, \omega^n$ are orthonormal),
(ii) $H = \sum_i b_i \otimes (\omega^i)^2$.

We may summarize (26) and (27) as meaning that there exists a special class of Darboux frames (x, e_i, e_μ) on X such that with respect to any one of these frames the ω^i satisfy (27). If we set

$$b_i = \sum_\mu b_i^\mu e_\mu ,$$

then by (26)

$$\sum_\mu b_i^\mu b_j^\mu = -1, \quad \text{if} \quad i \neq j,$$

and by (ii) of (27) we see that

$$\omega_i^\mu = b_i^\mu \omega^i \quad \text{(no sum)}.$$

Notice that these special frames are determined up to permutations, change of sign and orthogonal transformations of the normal part e_μ, (because the ω^i are uniquely determined, up to permutations and change of sign).

It is clear that the b_i span W (otherwise the $\hat{w}_i$ could not span $\hat{W}$), and so there is one generating relation

$$\sum_i \lambda_i b_i = 0.$$

As the λ_i are not all zero, we assume for the sake of argument that $\lambda_1 \neq 0$ and, multiplying by -1 if necessary, we may assume $\lambda_1 > 0$. Then, for $j > 1$, by (26)

$$0 = b_1 \cdot \sum_i \lambda_i b_i = \lambda_1 |b_1|^2 - \lambda_j - \sum_{i \neq 1, j} \lambda_i$$
$$0 = b_j \cdot \sum_i \lambda_i b_i = -\lambda_1 + \lambda_j |b_j|^2 - \sum_{i \neq 1, j} \lambda_i .$$

Subtracting, we have

$$\lambda_1(1 + |b_1|^2) = \lambda_j(1 + |b_j|^2)$$

from which we conclude that $\lambda_j > 0$ for all j. We thus may set $\lambda_i = 1/B_i > 0$ and normalize so as to have

$$(28) \qquad \begin{array}{ll} \text{(i)} & \sum_i b_i/B_i = 0 \\ \text{(ii)} & \sum_i 1/B_i = 1. \end{array}$$

Combining (26) and (28) we get

$$(29) \qquad B_i = 1 + |b_i|^2.$$

For future reference we note here that we have just shown that a unique set of relations (28) together with (29) hold for any set of vectors $b_1, \ldots, b_n \in W$ which satisfy (26) and span W.

Just as in Proposition (7) above we may compute the characteristic variety $\Xi_H \subset \mathbf{P}V^*$ of a 2nd fundamental form H given by (ii) in (27) with (28) holding.

(30) PROPOSITION: $\Xi_H = \Xi_{H,\mathbb{C}}$ consists of the $n(n-1)$ distinct real points

$$[\xi^{\pm}_{ij}] = [\sqrt{B_i}\,\omega_i \pm \sqrt{B_j}\,\omega_j] \qquad (i \neq j).$$

PROOF: As in the proof of (7) the condition that $[\xi] \in \Xi_H$ is, by 6.57, that there exist $w \in W$ and $0 \neq \eta \in V^*$ satisfying

$$w \cdot H = \xi\eta.$$

Now, for an H given by (ii) in (27), $w \cdot H$ has rank ≤ 2 if, and only if, for some i and j distinct we have

$$(31) \qquad w \cdot b_k = 0, \quad \forall\, k \neq i,j.$$

When $n = 2$ (31) imposes no conditions on $w \in W \cong \mathbb{R}^1$. In this case (28) says that

$$b_2 = -\frac{B_2}{B_1}\, b_1$$

and thus

$$H = \frac{b_1}{B_1} \otimes \left[B_1(\omega^1)^2 - B_2(\omega^2)^2\right].$$

Up to scalar multiple, for any $w \in W$, $w \cdot H$ is given by

$$B_1(\omega^1)^2 - B_2(\omega^2)^2 = (\sqrt{B_1}\,\omega^1 + \sqrt{B_2}\,\omega^2)(\sqrt{B_1}\,\omega^1 - \sqrt{B_2}\,\omega^2).$$

Thus the only possibility for ξ is either of the two factors on the right side of this equation.

When $n > 2$ then (31) implies that w lies in the orthogonal complement of $\mathrm{span}\{b_k : k \neq i, j\}$. But then, by reason of dimension, up to scalar multiples,

$$w = b_i - b_j.$$

Using (26) again we find that

$$\begin{aligned}(b_i - b_j)\cdot H &= (1 + b_i\cdot b_i)(\omega^i)^2 - (1 + b_j\cdot b_j)(\omega^j)^2 \\ &= B_i(\omega^i)^2 - B_j(\omega^j)^2\end{aligned}$$

by (29). Proposition (30) now follows immediately. □

As before, from this result together with (6.15) and (6.23) we may draw the following conclusion which is analogous to (10).

If the isometric embedding system for $\overline{X}^n \subset E^{2n-1}$ is involutive, then the local integral manifolds depend on $n(n - 1)$ functions of one variable. Moreover the system is hyperbolic.

Therefore, to prove theorem (24) we must show that the isometric embedding system is involutive. For this we will again use Cartan's test. We let

$$\mathcal{H} \subset W \times \ldots \times W \quad (n \text{ factors})$$

be the set of $(b_1, \ldots, b_n)$ satisfying

$$b_i \cdot b_j = -1, \quad i \neq j.$$

Remark that: a) $\mathcal{H}$ is a submanfold of dimension $n(n - 1)/2$, b) that each $(b_1, \ldots, b_n) \in \mathcal{H}$ satisfies a unique relation (i) in (28) with the normalization given by (ii) in (28), and c) that the argument leading to (29) gives that

$$B_i = 1 + b_i \cdot b_i .$$

On

$$M = \mathcal{F}(\overline{X}) \times \mathcal{F}(E^{2n-1}) \times \mathcal{H}$$

we consider the following variant of the isometric embedding system of Section 5

$$
(32) \quad
\begin{aligned}
&\text{(i)} && \omega^i - \overline{\omega}^i = 0 \\
&\text{(ii)} && \omega^\mu = 0 \\
&\text{(iii)} && \omega^i_j - \overline{\omega}^i_j = 0 \\
&\text{(iv)} && \omega^\mu_i - b^\mu_i \omega^i = 0 \quad \text{(no summation on i)} \\
&\text{(v)} && \bigwedge_i \omega^i \wedge \bigwedge_{\mu<\nu} \omega^\mu_\nu \neq 0
\end{aligned}
$$

where $b_i = (b^1_i, \ldots, b^{n-1}_i) = (b^\mu_i)$.

EXPLANATION: We are motivated by the summary following (27). Given a point $((y,\overline{e}_i),(x,e_i,e_\mu),(b_i))$ of M we define a 2nd fundamental form by

$$H = \sum_{\mu,i} b^\mu_i e_\mu \otimes (\overline{\omega}^i)^2.$$

As a consequence of (26) the Gauss equation

$$\gamma(H,H) = -\overline{ds}^2$$

are satisfied. The n-dimensional integral manifolds of (32) give partially framed isometric embeddings $\overline{X} \to E^{2n-1}$, where "partially framed" means that the normal frame is free to spin but the tangent frame is not. The reason for this is that for $b_i = (b^\mu_i)$, where the b^μ_i are components of a tensor in $\mathbb{R}^{n-1} \otimes \mathbb{R}^n$, the equations

$$b_i \cdot b_j = -1, \qquad i \neq j,$$

are invariant under rotations of $\mathbb{R}^{n-1}$ but not under rotations of $\mathbb{R}^n$. Put differently, an isometric embedding $\overline{X} \to \mathbf{E}^{2n-1}$ determines (up to permutation) a unique frame field on $\overline{X}$.

The Cauchy characteristic system of (32) is generated by the vector fields $\partial/\partial\omega^\mu_\nu$ $(\mu < \nu)$; i.e. is the Frobenius system given by

$$\{\overline{\omega}^i = 0,\ \omega^i = 0,\ \omega^\mu = 0,\ \omega^i_j = 0,\ \overline{\omega}^i_j = 0,\ \omega^\mu_j = 0\}.$$

As a consequence of the Gauss equations being satisfied, the structure equations of (32) are

$$\begin{array}{ll} \text{(i)} & d(\omega^i - \overline{\omega}^i) \equiv 0 \\ \text{(ii)} & d\omega^\mu \equiv 0 \\ \text{(iii)} & d(\omega^i_j - \overline{\omega}^i_j) \equiv 0 \\ \text{(iv)} & d(\omega^\mu_i - b^\mu_i\omega^i) \equiv -\sum_j (\delta_{ij}\beta^\mu_j - (b^\mu_i - b^\mu_j)\omega^i_j)\wedge\omega^j \end{array}$$

where

$$\beta^\mu_j = db^\mu_j + \sum_\nu \omega^\mu_\nu b^\nu_j$$

and where $\equiv$ denotes congruence modulo $\{\underline{I}\}$. Remark that in deriving (iv) we have, as usual, used the structure equations (23), and that in (iv) there is no summation on i. Remark also that since, for the reasons explained above, the ω^i_j do not appear in the independence condition (v) in (32), the terms

$$(b^\mu_i - b^\mu_j)\omega^i_j \wedge \omega^j$$

should not be considered as "part of the torsion" of (32). For the system (32) the forms ω^i_j and β^μ_j are the π^ϵ of the general theory in (4.1). Thus the ω^i_j will appear in the reduced tableau matrix (36) of (32), as they are not zero modulo $\{\underline{J}\}$.

It is now convenient to use vector-valued forms. Thus we set

$$\begin{aligned} \Omega_i &= {}^t(\omega^1_i, \ldots, \omega^{n-1}_i) \\ \beta_j &= {}^t(\beta^1_j, \ldots, \beta^{n-1}_j) = db_j + \omega b_j, \end{aligned}$$

where $\omega = (\omega^\mu_\nu)$, and we write (iv) as

$$(33) \quad d(\Omega_i - b_i\omega^i) \equiv -\sum_j (\delta_{ij}\beta_j - (b_i - b_j)\omega^i_j)\wedge\omega^j.$$

The fact that (i) in (28) is the only linear relation among the b_i's implies that for each fixed i the set of vectors $\{b_i - b_j | i \neq j\}$ is a basis for W. Thus, motivated by the term $-(b_i - b_j)\omega^i_j \wedge \omega^j$, we express β_i in terms of this basis,

$$\beta_i = \sum_j (b_i - b_j)\pi_{ij},$$

for unique 1-forms $\{\pi_{ij} | i \neq j\}$. Differentiation of (26) gives

$$b_i \cdot \beta_j + b_j \cdot \beta_i = 0 \quad \text{for} \quad i \neq j,$$

which, with (26) and (29), implies that

$$(34) \qquad B_j \pi_{ij} + B_i \pi_{ji} = 0, \quad i \neq j.$$

Recalling the defining equations for $\mathcal{H}$ and the β_i, it is clear that from among the $n(n - 1)$ π_{ij} $(i \neq j)$ there are at least $n(n - 1)/2$ independent forms; moreover, (34) gives $n(n - 1)/2$ independent relations. The conclusion is that (34) are the only relations among the π_{ij} $(i \neq j)$.

The structure equation (33) now is

$$(35) \quad d(\Omega_i - b_i \Omega^i) \equiv -\sum_j (b_i - b_j)(\pi_{ij} \wedge \omega^i - \omega^i_j \wedge \omega^j)$$

(no summation on i). The non-zero block of the reduced tableau matrix is the matrix of W-valued forms

$$(36)\quad \begin{vmatrix} \sum_j (b_i - b_j)\pi_{ij} & -(b_1 - b_2)\omega_2^1 \cdots & -(b_1 - b_n)\omega_n^1 \\ -(b_2 - b_1)\omega_1^2 & \sum_j (b_2 - b_j)\pi_{2j} & \cdots \\ \vdots & \vdots & \vdots \\ -(b_n - b_1)\omega_1^n & \cdots & \sum_j (b_n - b_j)\pi_{nj} \end{vmatrix}$$

The symbol relations are given by (34) plus $\omega_j^i + \omega_i^j = 0$. As in the preceeding case of a flat $X^n \subset E^{2n}$ the tableau matrix (36) is with respect to a non-generic flag. To compute the Cartan characters we must take a general linear combination of the columns of (36). When this is done we obtain for the first column

$$\begin{vmatrix} \xi_1 \sum_{j\neq 1} (b_1-b_j)\pi_{1j} - \xi_2(b_1-b_2)\omega_2^1 - \ldots - \xi_n(b_1-b_n)\omega_n^1 \\ -\xi_1(b_1-b_2)\omega_1^2 + \xi_2 \sum_{j\neq 2}(b_2-b_j)\pi_{2j} - \ldots - \xi_n(b_2-b_n)\omega_n^2 \\ \vdots \\ -\xi_1(b_1-b_n)\omega_1^n - \xi_2(b_2-b_n)\omega_2^n - \ldots + \xi_n \sum_{j\neq n}(b_n-b_j)\pi_{nj} \end{vmatrix}.$$

From the 1st entry we know the 1-forms

$$\xi_1\pi_{12} - \xi_2\omega_2^1,$$

and from the 2nd entry we know the 1-forms

$$-\xi_1\omega_2^1 + \xi_2\left(\frac{B_2}{B_1}\right)\pi_{12}.$$

Taking linear combinations we know π_{12} and ω_2^1. Continuing in this way we find that this vector contains the $n(n-1)/2 + n(n-1)/2 = n(n-1)$ independent forms

$$\pi_{ij},\ \omega_j^i \qquad (i < j).$$

From this we conclude that

$$s_1' = n(n-1),\ s_2' = \ldots = s_n' = 0,$$

in accordance with (30) and (6.9) (once we establish involutivity).

By Cartan's test, it remains to show that over each point there is an $n(n-1)$-dimensional family of integral elements. If $\{A_{ij} \mid i \neq j\}$ are arbitrary numbers, then the linear equations

$$\begin{aligned} \theta &= 0 \qquad \theta \in \underline{I} \\ \omega_j^i &= A_{ij}B_i\omega^i - A_{ji}B_j\omega^j \\ \pi_{ij} &= A_{ij}B_i\omega^j - A_{ji}B_j\omega^i \end{aligned}$$

define an $(n + (n-1)(n-2)/2)$-plane $E \in T_xM$ on which $\bigwedge_i \omega^i \wedge \bigwedge_{\mu<\nu} \omega_\nu^\mu \neq 0$ and which is clearly an integral element by (35). □

CHAPTER 8

EMBEDDING CAUCHY-RIEMANN STRUCTURES

All maps, manifolds and structures will be assumed to be of class C^∞. Of course when the Cartan-Kähler Theorem is applied for existence results everything must be assumed to be real analytic.

Let G be the Lie group

$$G = \{\begin{vmatrix} B & v \\ 0 & t \end{vmatrix} : B \in GL(n;\mathbb{C}),\ v \in \mathbb{C}^n,\ 0 \neq t \in \mathbb{R}\},$$

which is represented in $GL(2n+1;\mathbb{R})$ by

$$\begin{vmatrix} B & v \\ 0 & t \end{vmatrix} \to \begin{vmatrix} R(B) & -I(B) & R(v) \\ I(B) & R(B) & I(v) \\ 0 & 0 & t \end{vmatrix},$$

where $R(\cdot)$ and $I(\cdot)$ denote real part and imaginary part, respectively.

(1) DEFINITION: A <u>Cauchy-Riemann</u> (C-R) <u>structure</u> on a manifold X^{2n+1} is a G-structure; i.e., a reduction of the group of its tangent bundle to G. Locally it is given by 1-forms $\theta, \theta^1, \ldots, \theta^n$, where θ is real, and the θ^j are complex valued and they satisfy

(2) $$\theta\wedge\underset{j}{\theta^j}\wedge\bar{\theta}^j \neq 0.$$

Such forms are defined up to transformation by G. That is, if $\tilde{\theta}$, $\tilde{\theta}^1,\ldots,\tilde{\theta}^n$ also locally define the C-R structure then, on their common domain of definition U,

(3) $$\begin{vmatrix} \tilde{\theta}^j \\ \tilde{\theta} \end{vmatrix} = \begin{vmatrix} B & v \\ 0 & t \end{vmatrix} \begin{vmatrix} \theta^j \\ \theta \end{vmatrix}$$

for some smooth map

$$\begin{vmatrix} B & v \\ 0 & t \end{vmatrix} : U \to G.$$

The C-R structure is <u>integrable</u> if $d\theta$, $d\theta^j$ belong to the algebraic ideal generated by θ, θ^j. Thus, for an integrable structure

(4)
$$\text{i)}\quad d\theta = \theta^j\wedge\varphi_j + \theta\wedge\varphi$$
$$\text{ii)}\quad d\theta^j = -\varphi^j_k\wedge\theta^k + \theta\wedge\varphi^j,$$

for some 1-forms φ_j, φ^j_k, φ^j, φ, all of which are complex valued, except for φ which is real valued.

If X^{2n+1} and Y^{2n+1} have C-R structures, then a C-R isomorphism between them is a diffeomorphism

$f: X \to Y$ such that whenever $\tilde{\theta}, \tilde{\theta}^j$ locally defines the C-R structure on Y, then $f^*\tilde{\theta}, f^*\tilde{\theta}^j$ locally defines the C-R structure on X.

EXAMPLE: Real hypersurfaces $X^{2n+1} \subset \mathbb{C}^{n+1}$ form an important class of C-R manifolds. The C-R structure is naturally induced on X as follows. Let w, z^j denote complex coordinates on $\mathbb{C}^{n+1}$. Then X is locally defined by

$$(5) \qquad r(w, \overline{w}, z^j, \overline{z}^j) = 0,$$

where r is a real-valued function; without significant loss of generality we may assume that

$$\frac{\partial r}{\partial w} \neq 0.$$

The induced C-R structure on X is given locally by

$$\theta = \sqrt{-1}\, \partial r, \ \theta^j = dz^j, \quad \text{restricted to } X.$$

The induced structure is clearly integrable.

For an important special case of (5) let $g = (g_{ij})$ be any $n \times n$ hermitian matrix. Consider the real hyperquadric $Q(g)$ in $\mathbb{C}^{n+1}$ defined by (5) where

(6) $$r = \sqrt{-1}\,(w - \bar{w}) + g_{jk} z^j \bar{z}^k.$$

In this case we have

(7) $$\theta = -dw + \sqrt{-1}\, g_{jk} \bar{z}^k dz^j.$$

There is an alternate definition of C-R structure more frequently used by analysts. It is:

(8) A C-R structure on X^{2n+1} is an n-dimensional complex sub-bundle V of the complexified tangent space $\mathbb{C} \otimes TX$ for which $V \cap \bar{V}$ contains only the zero section. Such a structure is called integrable if

(9) $$[V,V] \subset V.$$

We shall indicate the equivalence of definitions (1) and (8). The equivalence of (4) and (9) will then be clear. Suppose that we have a C-R structure on X defined by the forms θ, θ^j, $j = 1,\ldots,n$, as in Definition (1). Then we have a linear isomorphism (at each point $x \in X$), where $\theta_x^{\perp} = \ker \theta_x \subset T_x X$,

(10) $$\psi : \theta_x^{\perp} \to \mathbb{C}^n,$$
$$v \to {}^t(\theta^1(v),\ldots,\theta^n(v))$$

The isomorphism ψ defines a complex structure J on $\theta^{\perp}$ by

$$\theta_x(J_x v) = \sqrt{-1}\ \theta_x(v).$$

Let

$$\begin{aligned}(11)\quad V_x &= \{\text{type } (0,1) \text{ vectors in } (\mathbb{C} \otimes \theta_x^{\perp}, J_x)\}\\ &= -\sqrt{-1} \text{ eigenspace of } J_x\\ &= \{v \in \mathbb{C} \otimes \theta_x^{\perp} : \theta^j(v) = 0,\ j = 1,\dots,n\}.\end{aligned}$$

The subbundle V then satisfies (8). Furthermore, if θ, θ^j satisfy (4), then V satisfies (9).

Conversely, given V of (8), locally there exist complex 1-forms $\theta^1,\dots,\theta^n$ in X which give a basis at each point x of $\overline{V}_x$. Furthermore, as $\overline{V}_x \oplus V_x$ has complex dimension $2n$, there exists a real 1-form θ, locally defined in X, such that

$$\overline{V}_x \oplus V_x = \text{kernel of } \theta \text{ in } \mathbb{C} \otimes T_xM.$$

Then $\theta, \theta^1,\dots,\theta^n$ are determined up to transformations (3) and thus define locally a C-R structure as in definition (1).

It is convenient to describe V in terms of a local basis, by which we mean locally defined sections (i.e. complex vector fields) $L_1, \ldots, L_n$ of V such that at each point x they give a basis over $\mathbb{C}$ of V_x. From (11) and the fact that $V_x \subset \mathbb{C} \otimes \theta_x^{\perp}$, we have

$$\theta^j(L_k) = 0 \quad \text{and} \quad \theta(L_k) = 0, \qquad j,k = 1,\ldots,n. \tag{12}$$

Reconsider example (5). As $X \subset \mathbb{C}^{n+1}$ we have $\mathbb{C} \otimes TX \subset \mathbb{C} \otimes T\mathbb{C}^{n+1}$. Hence a local basis $L_1, \ldots, L_n$ of V can be written

$$L_k = \xi_k \frac{\partial}{\partial w} + \eta_k \frac{\partial}{\partial \bar{w}} + \xi_k^j \frac{\partial}{\partial z^j} + \eta_k^j \frac{\partial}{\partial \bar{z}^j}, \tag{13}$$

for some complex-valued functions on X as coefficients. From (12) it follows that

$$\xi_k = \xi_k^j = 0.$$

By dimension count it follows that

$$V = (\mathbb{C} \otimes TX) \cap T^{0,1}\mathbb{C}^{n+1}. \tag{14}$$

As a consequence of (14) we see that if we regard the coordinate functions w, z^j restricted to X as complex-valued functions on X then these functions are solutions of the first order linear P.D.E. system

(15) $$L_k h = 0.$$

Furthermore, on X they satisfy

(16) $$dw \wedge dz^1 \wedge \ldots \wedge dz^n \neq 0.$$

The system (15) consists of generalized Cauchy-Riemann equations. If we set $L_k = X_k + \sqrt{-1}\, Y_k$, where X_k and Y_k are real vector fields in X, and we set $h = u + \sqrt{-1}\, v$, where u and v are real, then (15) becomes

(17) $$X_k(u) = Y_k(v)$$
$$Y_k(u) = -X_k(v).$$

These differ from the Cauchy-Riemann equations in that

$$[X_k, Y_k] \neq 0,$$

in general.

In the special case when $n = 1$ and X is the real hyperquadric $Q(g) \subset \mathbb{C}^2$ with $g = I_2 = 2 \times 2$ identity matrix, then (6) and (7) become

(18) $$r = -2v + |z|^2, \text{ and}$$

(19) $$\theta = -dw + \sqrt{-1}\, \bar{z} dz,$$

respectively, where $w = u + iv$ and $z = z^1$. Using (14) and (12) we find easily that

$$L_1 = \frac{\partial}{\partial \bar{z}} - \frac{\sqrt{-1}}{2} z \frac{\partial}{\partial u}, \tag{20}$$

which is the famous operator of Hans Lewy.

DEFINITION: The C-R embedding problem is the following: Given a manifold $\bar{X}^{2n+1}$ with an integrable C-R structure, consider the existence and uniqueness of embeddings

$$\bar{X} \overset{x}{\to} X \subset \mathbb{C}^{n+1} \tag{21}$$

such that x is C-R isomorphism between $\bar{X}$ and X with its induced C-R structure.

As for uniqueness it is an elementary exercise to verify that if F is a local biholomorphic transformation of $\mathbb{C}^{n+1}$ defined on a domain containing X, then F restricted to X is a C-R isomorphism between X and $F(X)$ with their induced structures. Conversely, in the real analytic case, if real hypersurfaces X and $Y \subset \mathbb{C}^{n+1}$ have isomorphic induced C-R structures then locally there exists a local biholomorphic transformation sending the portion of X in its domain onto a portion of Y. (cf. Cartan [4]).

A great deal of work has been done on the existence problem for (21). Before summarizing some of these results we define the Levi form of an integrable C-R structure.

By (i) in (4) and the fact that θ is real, it follows that

$$(22)\quad \begin{aligned} &\text{(i)}\quad d\theta \equiv -\sqrt{-1}\, g_{jk}\theta^j \wedge \bar{\theta}^k \mod \theta \\ &\text{(ii)}\quad g_{jk} = \bar{g}_{kj}. \end{aligned}$$

As is easily verified, the hermitian form

$$g_{jk}\theta^j \otimes \bar{\theta}^k$$

is, up to non-zero real factor, independent of the choice of θ and θ^j coming from transformations (3). This form is called the Levi form of the C-R structure. The structure is called strongly pseudo-convex if the Levi form is definite.

For the real hyperquadric $Q(g)$ of (6), the Levi form is

$$g_{jk}dz^j \otimes d\bar{z}^k \big|_{Q(g)}.$$

In 1972 Andreotti-Hill [1] showed the existence of

embeddings (21) in the real analytic case. In 1974 Boutet de Monvel [1] proved existence in the C^∞ category whenever $\bar{X}$ is compact of dimension ≥ 5 and the structure is strongly pesudoconvex. In the same year Nirenberg [1] gave an example of a C^∞ strongly pseudoconvex C-R structure on an $\bar{X}^3$ which cannot be even locally embedded in $\mathbb{C}^2$.

In order to explain Nirenberg's example, let $L_1,\ldots,L_n$ be a local basis of V for the given C-R structure on $\bar{X}^{2n+1}$. Suppose that h_a, $a = 0,1,\ldots,n$ are complex-valued functions on $\bar{X}$ which satisfy (15) and (16). Then

$$x = {}^t(h_0,\ldots,h_n):\bar{X} \to x(\bar{X}) = X \subset \mathbb{C}^{n+1}$$

defines a local embedding, with $h_o = w \circ x$, $h_j = z^j \circ x$. Now

$$\begin{aligned} dw(dx(L_k)) &= dx(L_k)(w) = L_k(h_o) = 0, \\ dz^j(dx(L_k)) &= dx(L_k)(z^j) = L_k(h_j) = 0, \end{aligned}$$

implies that $dx(V) \subset T^{0,1}\mathbb{C}^{n+1}$. Of course also $dx(V) \subset \mathbb{C} \otimes TX$. Hence, by (14), $dx(V)$ must define (by (8)) the induced C-R structure on X, and consequently x is a C-R isomorphism.

We see then that the existence problem for (21)

can be reformuated as: find $n + 1$ compex valued functions h_a, $a = 0,\dots,n + 1$ on $\bar{X}$ which satisfy (15) and (16). What Nirenberg did in [1] was to show that for a perturbation $\tilde{L}$ of the Lewy operator (20) the equation

$$\tilde{L}h = 0$$

has no non-constant solutions. Hence the integrable C-R structure on $\bar{X}^3$ defined by $\tilde{L}$ cannot be embedded.

Jacobowitz-Trèves [1] (which was used heavily for this chapter) generalized Nirenberg's procedure to show that the non-embeddable C^∞ C-R structures on $\bar{X}^3$ are "dense". Beyond that they showed that among the C^∞ integrable C-R structures on $\bar{X}^{2n+1}$, $n \geq 2$, whose Levi form has Lorentzien signature, the non-embeddable ones are dense. This contrasts with the 1982 result of Kuranishi [1] who proved the local embeddability of C^∞ strongly pseudo-convex structures on $\bar{X}^{2n+1}$ whenever $n \geq 4$.

What we shall do here is set up the C-R embedding problem in a way completely analogous to the way in which we have just studied the isometric embedding problem. To do this we consider $\mathbb{C}^{n+1}$ with the group of complex affine transformations

$\mathbb{C}A(n+1) = GL(n+1;\mathbb{C})\cdot\mathbb{C}^{n+1}$. The structure of this complex Lie group is easily seen from its representation in $GL(n+2;\mathbb{C})$ given by:

(23) $$\mathbb{C}A(n+1) = \{\begin{bmatrix} A & z \\ 0 & 1 \end{bmatrix} : A \in GL(n+1;\mathbb{C}), z \in \mathbb{C}^{n+1}\}.$$

The standard action of $GL(n+2;\mathbb{C})$ on $\mathbb{C}^{n+2}$ induce the action of $\mathbb{C}A(n+1)$ on

$$\mathbb{C}^{n+1} = \{\begin{bmatrix} w \\ 1 \end{bmatrix} : w \in \mathbb{C}^{n+1}\} \subseteqq \mathbb{C}^{n+2}.$$

If we define complex 1-forms

(24) $$\omega^a_b,\ \omega^a, \qquad 0 \leq a,b,c \leq n$$

on $\mathbb{C}A(n+1)$ by

(25) $$\begin{bmatrix} A & z \\ 0 & 1 \end{bmatrix}^{-1} d \begin{bmatrix} A & z \\ 0 & 1 \end{bmatrix} = \begin{bmatrix} \omega^a_b & \omega^a \\ 0 & 0 \end{bmatrix},$$

then they are a basis over $\mathbb{C}$ of the left-invariant holomorphic 1-forms on $\mathbb{C}A(n+1)$. Taking the exterior differential of (25) gives the Maurer-Cartan equations

$$(26)\qquad \begin{aligned} &\text{(i)} \quad d\omega^a = -\omega^a_b \wedge \omega^b \\ &\text{(ii)} \quad d\omega^a_b = -\omega^a_c \wedge \omega^c_b \end{aligned} \qquad 0 \le a,b,c \le n$$

(27) DEFINITION: Let $X^{2n+1} \subset \mathbb{C}^{n+1}$ be a real hypersurface. A local <u>C-R frame</u> in X is a map

$$e: U \to \mathbb{C}A(n+1)$$

on an open set $U \subset X$ such that

(i) $e(x) = (A,x)$ for every $x \in U$, where we write (A,x) for the element $\begin{bmatrix} A & x \\ 0 & 1 \end{bmatrix}$ of (23); and

(ii) $e^*\omega^0$ is real.

It is not hard to see that for any point in X there exists a C-R frame on some neighborhood of that point. (Cf. Jensen [1]). Furthermore, if e is a C-R frame in X and if we put

$$(28)\qquad \theta = e^*\omega^0,\ \theta^j = e^*\omega^j, \qquad j = 1,\dots,n,$$

then the corresponding forms defined by any other C-R frame in X will be related to those in (28) by the transformation (3). Thus (28) defines a C-R structure on X and it is in fact the induced C-R structure. (cf. loc. cit.).

We are thus led to the following formulation of the C-R embedding problem. Let $\overline{X}^{2n+1}$ have an integrable C-R structure locally defined by θ, θ^j which satisfy (4). On $M = \overline{X} \times \mathbb{C}A(n+1)$ we consider the P.D.S. (I,J) given locally by

$$\text{(29)} \qquad \begin{aligned} &\text{i)} \quad \omega^0 - \theta = 0 \\ &\text{ii)} \quad \omega^j - \theta^j = 0 \\ &\text{iii)} \quad \theta \wedge \underset{j}{\theta^j} \wedge \underset{j}{\overline{\theta}^j} \neq 0. \end{aligned}$$

There is a 1-1 correspondence between embeddings (21) and integral submanifolds of (29).

The structure equations of (29) are, by (4) and (26),

$$\text{(30)} \qquad \begin{aligned} d(\omega^0 - \theta) &\equiv -(\omega^0_j - \varphi_j) \wedge \theta_j - (\omega^0_0 - \varphi) \wedge \theta \\ & \qquad\qquad\qquad\qquad\qquad\qquad \operatorname{mod}\{\underline{I}\} \\ d(\omega^k - \theta^k) &\equiv -(\omega^k_j - \varphi^k_j) \wedge \theta^j - (\omega^k_0 - \varphi^k) \wedge \theta. \end{aligned}$$

If we take the real and imaginary parts of (30), and recall that the forms φ, φ^j, φ_j, φ^j_k of (4) are linear combinations of θ, θ^j, $\overline{\theta}^j$, we obtain the reduced tableau matrix of (29)

$$(31)\qquad \begin{array}{c|ccc|} & 0 & j & n+j \\ -1 & R(\omega_0^0) & R(\omega_j^0) & -I(\omega_j^0) \\ 0 & I(\omega_0^0) & I(\omega_j^0) & R(\omega_j^0) \\ k & R(\omega_0^k) & R(\omega_j^k) & -I(\omega_j^k) \\ n+k & I(\omega_0^k) & I(\omega_j^k) & R(\omega_j^k) \end{array}$$

where R and I denote real and imaginary parts, respectively.

If we label the rows and columns as indicated in (31), then we can write the symbol relations of (29) and (30) as

$$(32)\qquad \begin{aligned} \pi_j^{-1} &= \pi_{n+j}^{0}, & \pi_j^{0} &= -\pi_{n+j}^{-1} \\ \pi_j^{k} &= \pi_{n+j}^{n+k}, & \pi_j^{n+k} &= -\pi_{n+j}^{k}. \end{aligned}$$

An easy count yields: The reduced characters of (29) are:

$$(33)\qquad s_1' = \ldots = s_{n+1}' = 2n+2, \quad s_{n+2}' = \ldots = s_{2n+1}' = 0.$$

(34) PROPOSITION: The characteristic variety of (29) at $x \in \overline{X}$ consists of a single point,

$$\text{(i)}\qquad \Xi_x = [\theta_{(x)}] \in P(J_x/I_x).$$

The complex characteristic variety of (29) is

$$(\text{ii})\ \Xi_{x,\mathbb{C}} \cong (n+1)(\mathbb{C}P^n)^+ + (n+1)(\mathbb{C}P^n)^- \subset \mathbb{C}P^{2n},$$

where for $\epsilon = +$ or $-$,

$$(\mathbb{C}P^n)^\epsilon = \{[\xi_o, \xi_j, \xi_{n+j}] : \xi_{n+j} = \epsilon\sqrt{-1}\ \xi_j,\ \xi_j \in \mathbb{C}\}.$$

More explicitly, set theoretically,

$$(\text{iii})\quad \Xi_{x,\mathbb{C}} = \{[\xi_o\theta + \xi_j\theta^j]\} \cup \{[\xi_o\theta + \xi_j\bar{\theta}^j]\} \subset P(\mathbb{C} \otimes J_x/I_x).$$

Thus the dimension of $\Xi_{x,\mathbb{C}}$ is n and its scheme-theoretic degree is $2n + 2$, as predicted by (33) and (6.15) once we show that (29) is involutive.

PROOF: Let $\xi = (\xi_o, \xi_j, \xi_{n+j})$ and $t = {}^t(t^{-1}, t^0, t^j, t^{n+j})$. From (32) the symbol map is (i.e. substitute $t^\alpha\xi_\rho$ in place of π^α_ρ) at x:

$$(35)\qquad \sigma_\xi(x)t = \begin{vmatrix} t^{-1}\xi_j - t^o\xi_{n+j} \\ t^o\xi_j + t^{-1}\xi_{n+j} \\ t^k\xi_j - t^{n+k}\xi_{n+j} \\ t^{n+k}\xi_j + t^k\xi_{n+j} \end{vmatrix}.$$

Hence the equations of $\Xi_{x,\mathbb{C}}$ are the $(2n + 2)\times(2n + 2)$ minors of $\sigma_\xi(x)$. To determine the set-theoretic locus of these equations, suppose that $\sigma_\xi(x)t = 0$ and that $t^{-1} \neq 0$ and $t^0 \neq 0$. Then we have

$$t^{-1} = t^0 \frac{\xi_{n+j}}{\xi_j} = -t^0 \frac{\xi_j}{\xi_{n+j}}, \quad \text{for all } j.$$

Thus, for all j

$$\xi_j^2 = -\xi_{n+j}^2,$$

which implies that

$$\xi_{n+j} = \epsilon_j\sqrt{-1}\,\xi_j, \quad \text{for all } j,$$

where each $\epsilon_j = \pm 1$. However, we also have

$$\frac{t^{-1}}{t^0} = \frac{\xi_{n+j}}{\xi_j} = \epsilon_j\sqrt{-1}, \quad \text{for all } j,$$

which implies that the ϵ_j are all equal, to $\epsilon = \pm 1$ say. An analysis of the cases where $t^{-1} = 0$ or $t^0 = 0$ yields no additional points in the locus. Hence the set theoretic locus of the equations of $\Xi_{x,\mathbb{C}}$ is

$$(\mathbb{C}P^n)^+ \cup (\mathbb{C}P^n)^-,$$

where $(\mathbb{C}P^n)^{\pm} \subset \mathbb{C}P^{2n}$ are defined in (34). This proves (34) (ii). One obtains (iii) from (ii) by using the basis $\theta, R(\theta^j), I(\theta^j), j = 1,\dots,n$ of J_x/I_x. Finally, (i) follows from (iii), as the points $[\xi_0\theta + \xi_j\theta^j]$ or $[\xi_0\theta + \xi_j\overline{\theta}^j]$ are real if, and only if, $\xi_j = 0$ and ξ_0 is real. □

(36) THEOREM: The C-R embedding system (29) is involutive. Its general solution depends on $2n + 2$ real functions of $n + 1$ real variables.

PROOF: We do this by applying Cartan's Test, which means we must compute the dimension t of the linear variety of integral elements of (29) at a point of M. The equations of the integral elements are (i) and (ii) of (29) plus

$$(37)\qquad \begin{aligned} \omega_j^0 - \varphi_j &= A_{jk}\theta^k + B_{jk}\overline{\theta}^k + C_j\theta \\ \omega_0^0 - \varphi &= D_k\theta^k + E_k\overline{\theta}^k + F\theta \\ \omega_0^j - \varphi^j &= H_k^j\theta^k + I_k^j\overline{\theta}^k + J^j\theta \\ \omega_k^j - \varphi_k^j &= N_{k\ell}^j\theta^\ell + Q_{k\ell}^j\overline{\theta}^\ell + P_k^j\theta \end{aligned}$$

where the coefficients are complex numbers which satisfy the relations obtained by substituting (37) into the structure equations (30). These relations are

$$(38) \qquad \begin{aligned} &A_{jk} = A_{kj}, \ B_{jk} = 0, \ C_j = D_j, \ E_j = 0 \\ &N^j_{k\ell} = N^j_{\ell k}, \ Q^j_{k\ell} = 0, \ P^j_k = H^j_k, \ I^j_k = 0. \end{aligned}$$

Counting the number of independent real coefficients in (37) which satisfy the relations (38) we obtain

$$(39) \qquad t = (n + 1)^2(n + 2)$$

which is equal to $s'_1 + 2s'_2 + \ldots + (n + 1)s'_{n+1}$ by (33).

□

REFERENCES

Adams, J.F.; Lax, P.D.; Phillips, R.S.

[1] On matrices whose real linear combinations are non-singular, Proc. Amer. Math. Soc. 16 (1965), 318-322.

Allendoerfer, C.B.

[1] Rigidity for spaces of class greater than one, Amer. J. Math. 61 (1939), 633-644.

Aminov, Yu. A.

[1] Isometric immersions of domains of a three-dimensional Lobachevskii space in a five-dimensional Euclidean space, and motion of a rigid body, Math. USSR-Sb 50 (1985), no. 1, 11-30.

Andreotti, A.; Hill, C.D.

[1] Complex characteristic coordinates and tangential Cauchy-Riemann equations, Annali della Scuola Normale Superiore di Pisa, 26 (1972), 299-324.

Boutet de Monvel, L.

[1] Intégration des équations de Cauchy-Riemann induites formeles, Seminaire Goulaouic-Lions-Schwartz (1974-75).

Berger, E.; Bryant, R.; Griffiths, P.

[1] Some isometric embedding and ridigity results for Riemannian manifolds, Proc. Nat. Acad. Sci. 78 (1981), 4657-4660.

[2] The Gauss equations and rigidity of isometric embeddings, Duke Math. J. 50 (1983), 803-892.

Bryant, R.; Chern, S.S.; Gardner, R.; Goldschmidt, H.; Griffiths, P.; Yang, D. = Bryant, et al.

[1] Essays on exterior differential systems, in preparation.

Bryant, R.; Griffiths, P.; Yang, D.

[1] Characteristics and existence of isometric embeddings, Duke Math. J. 50(1983), 893-994.

Cartan, E.

[1] Les systems diffèrentielles extérieurs et leurs applications géométriques, Hermann (1945), Paris.

[2] Sur la possibilité de plonger un espace riemannien donné dans un espace euclidien, Ann. Soc. Pol. Math 6 (1927), 1-7. Or Oeuvre Partie III, v. 2, 1091-1098.

[3] Sur les variétés de courbure constante d'un espace euclidien ou non euclidien, Bull. Soc. Math. France, t. 47 (1919), 125-160; and t. 48 (1920), 32-208. Or Oeuvre Partie III, v. 1, 321-432.

[4] Sur la géométrie pseudo-conforme des hypersurfaces de deux variables complexes, Ann. Math. Pura Appl. (4) 11 (1932), 17-90. Or Oeuvres Partie II, v. 2, 1231-1304.

Chern, S.S.

[1] A proof of the uniqueness of Minkowski's problem for convex surfaces, Amer. J. Math 79 (1957), 949-950.

Chern S.S.; Kuiper, N.

[1] Some theorems on the isometric imbedding of compact Riemann manifolds in Euclidean space, Ann. of Math. 56 (1952), 422-430.

DoCarmo, M.

[1] Differential Geometry of Curves and Surfaces, Prentice-Hall, Englewood Cliffs, 1976.

Efimov, N.V.

[1] Generation of singularities on surfaces of negative curvature, Math. Sbornik 64 (106) (1964), 286-320.

Greene, R.

[1] Isometric embeddings of Riemannian and pseudo-Riemannian manifolds, Memoirs of the Amer. Math. Soc. 97, (1970).

Gromov, M.L.

[1] Differential relations, to appear.

Gromov, M.L.; Rokhlin, V.A.

[1] Embeddings and immersions in Riemannian geometry, Russian Math Surveys 25 (1970), 1-57.

Hamilton, R.S.

[1] The inverse function theorem of Nash and Moser, Bull. Amer. Math. Soc. (N.S.) 7 (1982), 65-222.

Jacobowitz, H.

[1] Local isometric embeddings, Seminar on Differential Geometry, S.S. Yau editor, Annals of Math Study's 102 (1982), Princeton, New Jersey.

Jacobowitz, H.; Trèves, F.

[1] Nonrealizable CR structures, Invent. Math. 66 (1982), no. 2, 231-249.

Jensen, G.

[1] Projective deformation and biholomorphic equivalence of real hypersurfaces, Ann. Glob. Analysis and Geom. 1 (1983), no. 1, 1-34.

John, F.

[1] Partial Differential Equations, 4th edition, Springer-Verlag, New York, 1982.

Kaneda, E.; Tanaka, N.

[1] Rigidity for isometric embeddings, J. Math. Kyoto Univ. 18 (1978), 1-70.

Kobayashi, S.; Nomizu, K.

[1] Foundations of Differential Geometry, Vol. II, John Wiley & Sons, New York, 1969.

Kuranishi, M.

[1] Strongly pseudoconvex CR structures over small balls Part III. An embedding theorem. Annals of Math 116 (1982), 249-330.

Lin, Chang-Shou.,

[1] The local isometric embedding problem in $\mathbb{R}^3$ of two dimensional Riemannian manifolds with Gaussian curvature changing sign nicely. Thesis, NYU, New York (1983).

Moore. J.D.

[1] Isometric immersions of space forms in space forms, Pac. J. of Math. 40 (1972), 157-166.

Nash, J.

[1] The embedding problem for Riemannian manifolds, Ann. of Math. 63 (1956), 20-64.

Nirenberg, L.

[1] On a problem of Hans Lewy, Uspeki Mat. Nauk 292 (176) (1974), 241-251.

Spivak, M.

[1] A Comprehensive Introduction to Differential Geometry, Vol. V, Publish or Perish, Inc. Boston, 1975.

Steiner, S.; Teufel, E.; Vilms, J.

[1] On the Guass equation of an isometric immersion, Duke Math. J. 51 (1984), no. 2, 421-430.

Tanaka, N.

[1] Rigidity for elliptic isometric embeddings, Nagoya Math. J. 51 (1973), 137-160.

Tenenblat, K.

[1] A rigidity theorem for three-dimensional submanifolds in Euclidean six space, J. Diff. Geom. 14 (1979), 187-203.

Vilms, J.

[1] Local isometric imbedding of Riemannian n-manifolds into Eucliden (n+1)-space, J. Diff. Geom. 12 (1977), 197-202.

[2] Factorization of curvature operators, Trans. Amer. Math. Soc. 260 (1980), 595-605.

Yang, D.

[1] Involutive hyperbolic differential systems, Thesis, Harvard Univ., Cambridge, Mass (1982).

INDEX

Library of Congress Cataloging-in-Publication Data

Griffiths, Phillip, 1938-
Differential systems and isometric embeddings.

(Annals of mathematics studies ; 114)
''William H. Roever lectures in geometry''—Foreword.
Bibliography: p.
1. Exterior differential systems. 2. Differential equations, Partial. 3. Embeddings (Mathematics) 4. Riemannian manifolds. I. Jensen, Gary R., 1941- . II. Title. III. Title: William H. Roever lectures in geometry. IV. Series: Annals of mathematics studies ; no. 114.
QA649.G825 1987 515.3′53 86-43134
ISBN 0-691-08429-7
ISBN 0-691-08430-0 (pbk.)

Phillip A. Griffiths is Provost of Duke University. Gary R. Jensen is Professor of Mathematics at Washington University in St. Louis

Lightning Source UK Ltd.
Milton Keynes UK
UKHW010653040322
399558UK00001B/65

9 780691 084305